MOVIE MATH MANIA

AIMS EDUCATION FOUNDATION PRESENTS:
"MOVIE MATH MANIA"
WRITTEN BY SHELDON ERICKSON
ILLUSTRATED BY BROC HEASLEY   EDITED BY BETTY CORDEL
DESKTOP PUBLISHING BY TANYA ADAMS

This book contains materials developed by the AIMS Education Foundation. **AIMS** (**A**ctivities Integrating **M**athematics and **S**cience) began in 1981 with a grant from the National Science Foundation. The non-profit AIMS Education Foundation publishes hands-on instructional materials (books and the monthly magazine) that integrate curricular disciplines such as mathematics, science, language arts, and social studies. The Foundation sponsors a national program of professional development through which educators may gain both an understanding of the AIMS philosophy and expertise in teaching by integrated, hands-on methods.

ISBN **978-1-932093-02-5**
Printed in the United States of America

I Hear and I Forget,
I See and I Remember,
I Do and I Understand

Chinese Proverb

MOVIE MATH MANIA

Table of Contents

INTRODUCTION

Today's student is a part of a very visual culture; one that is infused with media. A child growing up today has spent more time watching movies and videos than reading. Movies are in some ways more a part of a student's life than actual experience.

In order to make math meaningful, it must connect with the child's world. Looking at popular movies provides contexts that connect with the background experience of today's students. This book explores possible ways to use the content of middle school math in the context of movies. The movies provide the motivation and purpose for studying the math. The math provides the understanding of what is involved in making a movie or the experiences portrayed in the movies.

In most cases, it is not an appropriate use of math instructional time to watch a whole feature length movie. In some cases, parts of the movies are not even appropriate for young children due to violence or language; however, if a teacher acquires the video and shows the appropriate clips, a context and motivation have been set up for doing the math. Many students will have already seen the movie or similar situations and will relate the viewed clip to the whole story.

The activities in this book provide just a sample of how movies might be used to enhance math instruction. It is not meant to be an exhaustive list of appropriate movies or investigations. The author hopes that these suggestions might spark teachers' creativity and that many other appropriate movies and applications will be developed. Hopefully math teachers will begin to see popular culture as a tool for providing motivation and meaning to students' learning.

Scaling Films

There is a whole genre of films based on miniature people. Disney created a sequence of films including *Honey, I Shrunk the Kids, Honey, We Shrunk Ourselves,* and *Honey, I Blew up the Kids.* Other movies are based on book series including *Indian in the Cupboard,* and *The Borrowers.* The two latter choices provide a literary component that can provide very rich integrated experiences.

The actors in these types of movies are made to look small using a combination of techniques. One technique is to film the actor in front of a vivid green backdrop. The backdrop is removed electronically and the actor is superimposed over a scaled background that makes the actor appear to be much smaller than in real life. This technique is used in the television weather forecast where the meteorologist is walking in front of a large satellite image of the weather. This is the least expensive technique, but does not allow the actor to interact with props.

To make the miniaturization of actors seem more realistic, objects are enlarged. As the actor stands near or interacts with these enlarged props, he/she appears smaller than they are. To make it seem real, the producer spends a good deal of time and expense making sure the props maintain a very realistic appearance in shape, color, texture, and relative weight. As actors interact with scaled props against an enlarged background it seems very plausible that they have been miniaturized.

A great deal of proportional reasoning develops as students consider the scaling in these movies. Looking at and measuring several scenes in which the scaling of props or background has been used allows students to see what mathematics was necessary in preparing the movie. This experience also provides the students with the opportunity to see how producers are not as concerned about accuracy as they are with visual effects.

The opportunity of constructing a scaled prop allows students the chance to apply their understanding and to recognize all the dimensional effects of enlargement. With a class set of props, the students are ready to make their own movie to add to the list.

Topic
Proportional reasoning—scaling

Key Question
How can you determine what scale the movie producers used to make the props for this movie?

Focus
Students will
1. make measurements of scenes from movies to determine if the props remain true to scale,
2. use these measurements to help them develop a better understanding of proportional reasoning and apply it to determine if objects are scaled proportionately.

Guiding Documents
Project 2061 Benchmark
- *Estimate distances and travel times from maps and the actual size of objects from scale drawings.*

NCTM Standards 2000*
- *Understand and use ratios and proportions to represent quantitative relationships*
- *Develop, analyze, and explain methods for solving problems involving proportions, such as scaling and finding equivalent ratios*
- *Solve problems involving scale factors, using ratio and proportion*

Math
Proportional reasoning
    scaling
Measurement
    length

Integrated Processes
Observing
Comparing and contrasting
Generalizing
Applying

Materials
Rulers
Student pages
Light bulbs
Wooden matches
Paper clips
Credit cards (copies)
Lego® bricks (2 x 4 studs)
Straight pins
Spoons

Background Information
    In movies in which humans appear miniature, a scale factor for props is used to determine how small the actors will appear. All props are enlarged by this factor to make the apparent size of the actors consistent. The scale of enlargement may be ignored in order to enliven the story or make the scenes more plausible. In most movies, there is some reference explaining the scale between the scaled prop and the apparent size of the actor. By measuring scenes with actors and props, it is possible to determine how well the scale is maintained.

    In the movie, *The Borrowers*, this scale is made directly when Arrietty, the girl, stands 4.5 inches tall by a tape measure. The height of the rest of the Clock family can be determined by comparing their picture heights to Arrietty's. By multiplying the ratio of the picture heights by 4.5 you would get the movie height in inches. Any prop can be compared to the apparent height of any member of the family to determine the apparent size of each prop. Seeing how the apparent size of the prop compares to the actual object can check the consistency of the scale.

Character	Picture Height	Ratio	Apparent Height
Arrietty	2.125		4.5
Spud	2.375	$\frac{2.375}{2.125} = \frac{S}{4.5}$	5.03
Peagreen	2.0	$\frac{2.0}{2.125} = \frac{P}{4.5}$	4.24
Ocious	2.5	$\frac{2.5}{2.125} = \frac{O}{4.5}$	5.29
Homily	2.375	$\frac{2.375}{2.125} = \frac{H}{4.5}$	5.03

All measurements are in inches

Character / Object	Picture Ratio	=	Apparent Ratio	Object Size
Peagreen / lightbulb	$\dfrac{2.215}{2.0}$	=	$\dfrac{4.24}{L}$	L = 3.98
Spud / match	$\dfrac{2.625}{1.25}$	=	$\dfrac{5.03}{M}$	M = 2.4
Ocious / paper clip	$\dfrac{2.5}{0.625}$	=	$\dfrac{5.29}{P}$	P = 1.32
Homily / credit card	$\dfrac{2.625}{2.0}$	=	$\dfrac{5.03}{C}$	C = 3.83

All measurements are in inches

In the movie *Honey, I Shrunk the Kids,* Nick states he is one-fourth inch high, 64 feet (768 inches) from the house, which is like 3.2 miles (5280 • 3.2 = 16896 inches). This sets up a proportion that allows you to determine how tall the movie producers consider Nick.

	Real Ratio	=	Apparent Ratio	Character Size
Character / Distance	$\dfrac{0.25 \text{ inches}}{768 \text{ inches}}$	=	$\dfrac{N \text{ feet}}{16896 \text{ feet}}$	N = 5.5 feet

Five and a half feet tall seems too big for a ten year old kid, but that is what the producer is figuring. Was this scale consistent?

Character / Object	Picture Ratio	=	Apparent Ratio	Object Size
Character / LEGO® brick	$\dfrac{1.625}{2.625}$	=	$\dfrac{0.25}{L}$	L = 0.4
Character / pin	$\dfrac{0.75}{3.0}$	=	$\dfrac{0.25}{P}$	P = 1.0
Character / spoon	$\dfrac{0.5}{3.0}$	=	$\dfrac{0.25}{S}$	S = 1.5

All measurements are in inches

As students are developing proportional reasoning skills, they may not set up two ratios as a proportion as shown. Many will consider the situation and do several one step calculations that provide a meaningful understanding.

Many students determine Ocious Clock's height by first dividing his picture height (2.5 inches) by Arriety's picture height (2.125 inches) and saying that he is 1.18 times taller than her. Then they multiply Arriety's apparent height (4.5 inches) by this factor to get Ocious's apparent height (5.31 inches).

Some students will be even more concrete. When they measure Nick's height on the spoon as .5 inch, they will mark off half-inch segments across the spoon (3 inches) to see that six Nicks can lie across the spoon. They reason that if Nick is a quarter inch long, the spoon must be 1.5 inches long.

These are very important steps in the development of students' understanding. Teachers should encourage students to explain their understanding and show the students how the proportion formalizes their thinking.

Management

1. The teacher might want to show clips from movies that use miniaturization to provide a better understanding of the context and the techniques used to make these types of movies.

2. The students should have some experience with proportional reasoning before tackling the more difficult numbers this activity generates. A developmentally appropriate activity to do before this investigation is *Picturing Proportions* from *Proportional Reasoning,* an AIMS publication.

3. Illustrations of scenes from two movies have been provided to give more experience with the concepts of scaling. If a teacher plans to use both, the students will find the scenes from *Honey, I Shrunk the Kids* more straightforward for the first experience.

4. Have a sample of each object for each group if possible. It provides a concrete object that many students can use to compare their predictions when determining how to calculate the object's apparent size.

Procedure

1. Show some of the illustrations of actors interacting with props. Then pose the *Key Question*.

2. Distribute the illustrations of the scenes and rulers. Have the students determine the ratio that communicates the scale used for the props from the first frame.

3. Allow the students to grapple with how to use this ratio proportionately to determine how tall Nick should have been before he was shrunk or how tall each member of the Clock family is.

4. Have students share their methods for solutions for this problem. Assess that they have correctly applied proportional ideas to get their solutions.

5. Distribute the objects represented by the props in the illustrations and have students use proportional reasoning skills to determine the apparent size of the objects. Allow the students to check their solutions and see how well the scale was maintained with the actual objects.

6. Have students share the variety of ways they thought about determining the apparent size of the objects. As a class develop a proportion to represent the situation and have students identify how their thinking can be seen in the proportion.

7. Using the actual objects, have students discuss how well they think the movie props maintained a consistent scale. Have the students suggest reasons the producers may have allowed for any inconsistencies.

Discussion

1. How tall was Nick before he was shrunk? [5.5 ft.]

2. How tall is each of the Clock family members? (See *Background Information.*)

3. How did you determine the size an object from the picture and scale? (Answers will vary. See *Background Information.*)

4. What proportion could represent this situation? (See *Background Information.*)

5. Can you see the arithmetic steps you did for your solution in the proportion? Explain.

6. How well were the props made to the same scale? (Accuracy will vary.)

7. What are some reasons producers might not keep things to an exact scale?

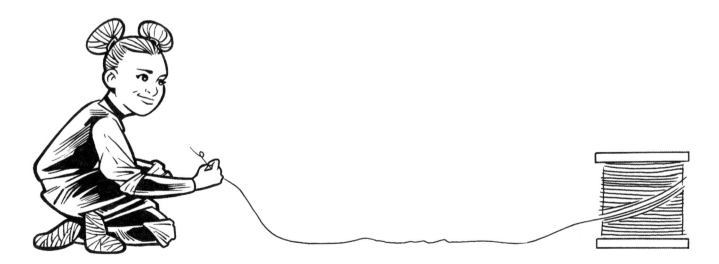

From Honey, I Shrunk the Kids

From *The Borrowers*

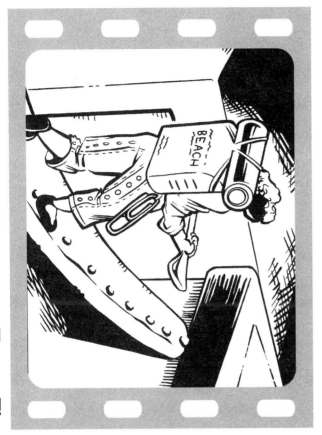

MAKING MOVIE PROPS

Topic
Proportional reasoning—scaling

Key Question
How can you make a prop for a movie so that it looks like an actor your size has been shrunk to the size of your action figure?

Focus
Students will:
1. use their heights and the height of an action figure to establish a scale, and
2. use that scale to construct an enlarged object they can interact with that makes them appear to be the size of the action figure.

Guiding Documents
Project 2061 Benchmarks
- *The scale chosen for a graph or drawing makes a big difference in how useful it is.*
- *Scale drawings show shapes and compare locations of things very different in size.*
- *Estimate distances and travel times from maps and the actual size of objects from scale drawings.*

NCTM Standards 2000*
- *Understand and use ratios and proportions to represent quantitative relationships*
- *Develop, analyze, and explain methods for solving problems involving proportions, such as scaling and finding equivalent ratios*
- *Solve problems involving scale factors, using ratio and proportion*
- *Select and apply techniques and tools to accurately find length, area, volume, and angle measures to appropriate levels of precision*

Math
Proportional reasoning
    scaling
Measuring
    length

Integrated Processes
Observing
Collecting and recording data
Applying
Generalizing

Materials
Action figures
Rulers or tape measures
Construction materials and tools (see *Management*)

Background Information
Many popular movies are based on the idea of humans "shrunk" in proportion to their surroundings. In *Honey, I Shrunk the Kids,* children are shrunk to be one-fourth inch high. In *Indian in the Cupboard*, miniature figures come to life in a normal world. *The Borrowers* has miniature human-type characters living with full-size humans. The concept of shrinking has captivated our imaginations for generations. Movies have capitalized on this interest by using two "tricks" to make human actors look shrunken.

The first trick is to have the actors play out a scene in front of a blank background. A second film is shot of the full-size background in which the actors will appear. The film of the actors is then overlaid on top of the full-size background. The combination of the two films tricks the brain into assuming that the humans are smaller than the background that surrounds them. This method is the easiest to create, but it does not allow the actors to interact with the objects that surround them.

The second method is to make props by enlarging the objects with which the actors will interact. The scene is shot with the actors working on stage with the props. This method tricks the brain into assuming that the actors are small because they are the size of small common objects.

The films mentioned use a combination of the tricks, selecting the one that is most appropriate for each scene. To be convincing, a scale must be chosen and used throughout the film. Before filming can begin, the shrunken size of the actors must be determined. The background footage must make things appear as many times taller than the shrunken actor as the scale defines. The scale factor of the actor to his or her shrunken counterpart is used to multiply the size of the objects when constructing props.

In *True Scales* students determined the scale that producers used to make their movies. Now they will be making the prop using the scale of themselves to an action figure.

Many students who are not secure with their proportional-reasoning skills will think through their solutions and check them in a very concrete manner. They will thusly confirm their mathematical computations with their more intuitive methods.

To establish scale, many students see how many action figures tall the human actor will be and use that as their scale. Other students will place the object they will make into a prop by the action figure to make direct comparison. If the object is as tall as the action figure's waist, they make the prop as tall as the actor's waist. These direct comparison methods are mimicked in the arithmetic. A person's height is divided into action figure's height to determine how many times bigger (scale) the actor is than the action figure. It is proportionally reasoned that if an object is half the waist height of an action figure, its enlarged prop should be half the waist height of the actor. One of the goals of this lesson is to help students recognize that what they understand intuitively can be done more accurately mathematically.

Management

1. The teacher should ask students to bring the action figures to class before introducing the project. A selected figure will be used to establish a scale. (A teacher may choose to supply the same size action figures. This will maintain a more consistent scale for all the props so a class-wide video can be made.)

2. This activity works well as a small group project with two or three students. A class period might be provided in class to introduce the project and have students get started. A second period should be used to present the projects at the end. A week between the introduction and the presentation should be sufficient for students to complete the construction of the props on their own time.

3. Students will choose various methods of constructing their props considering size and appearance. The teacher may choose to make students responsible for their own construction materials or to provide some basic construction materials and tools, such as cardboard, card stock, butcher paper, construction paper, scissors, glue, tape, paint, etc.

4. Optional: Have a camera available to take photographs of each group's object with their action figure and the "actor" with the prop. This provides a nice way to assess accuracy of scale and provide for a long-term display.

Procedure

1. Ask students to bring an action figure to school for building a movie prop.

2. Have the class make a list of movies that have scenes with miniature people. Discuss how these films were made to make the people appear to be that small. Discuss the *Key Question*.

3. Direct each group to measure and record the height of the action figure and the student chosen to be the "actor."

4. Using the height measurements, have each group determine the appropriate scale factor.

5. Inform students that each group is responsible for finding an object to use as a prop. Encourage them to be thoughtful about their selection, considering the size the object will need to be in order to be proportional to the human actor.

6. Have the groups measure and record the length, width, height, and any other critical dimensions of the object they have chosen.

7. Using the scale and data, direct each group to determine the dimensions of the prop.

8. Have each group build a prop to the correct dimensions.

9. Inform students that they will need to bring in the action figure, object, prop, and actor to present their solution. Encourage the class to determine whether the proportions were maintained by comparing the figure and the object to the actor and the prop.

Discussion

1. How do you use the height of the figure and the actor to determine the scale for your prop? [actor's height: figure's height]

2. What reasons did you have for choosing the object for your prop?

3. How can you determine a scale has been maintained by looking at the object, figure, prop, and actor? [The object-to-figure comparison looks the same as prop-to-actor comparison.]

4. Why can't all the props be used in the same movie? [Each is based on a different scale.]

Extension

Have the students write scripts for short sequences of a movie in which the actor interacts with the prop. These can be videotaped.

*   Reprinted with permission from *Principles and Standards for School Mathematics,* 2000 by the National Council of Teachers of Mathematics. All rights reserved.

MAKING MOVIE PROPS

Make a prop for a movie so that it looks like an actor your size has been shrunk to the size of your action figure.

Measure the actor's height and the action figure's height.

Actor's Name: _____ Actor's Height: _____

Figure's Name: _____ Figure's Height: _____

Show how you determined how many times bigger the actor is than the figure.

Find and measure the object you are going to enlarge into a prop.

Description of object:

Critical dimensions:
        Length: _____ Width: _____ Height:_____

Show how you determined how big to make the prop.

Prop dimensions:

Length: _____ Width:_____ Height:_____

CAST AWAY

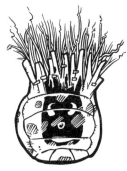

Chuck Noland is a systems engineer for Federal Express who works and lives by the clock. As a high strung, time oriented, aggressive businessman, he finds himself stranded on a deserted island with little hope of rescue after his plane crashes. Becoming extremely resourceful and creative with recovered materials, he survives and escapes from the island.

Removed from his dependency on modern technology, Chuck often demonstrates his skill of estimation, which was developed in business dealings.

In one scene he estimates where he is located and the area of the search required to find him. Although he concludes correctly that he would not be found, he provides an excellent example of the process and skills required in estimation.

To get off the island, Chuck builds a raft and estimates the rope and time required to make a timely escape. Again Chuck models the thinking skills required in estimation and provides a context that is of interest to students.

Cast Away received a PG-13 rating for some intense images and action sequences involving the plane crash. The two scenes dealing with the estimation have no objectionable content.

HERE IN THE WORLD?

Topic
Rates and scale maps

Key Question
If you only had the information from Chuck Noland's monologue, where would you search for him?

Learning Goals
Students will:
1. learn to determine distance given rate and time, and
2. learn to determine position from distance and directional clues.

Guiding Documents
Project 2061 Benchmarks
- *Mathematics is helpful in almost every kind of human endeavor—from laying bricks to prescribing medicine or drawing a face. In particular, mathematics has contributed to progress in science and technology for thousands of years and still continues to do so.*
- *Estimate distances and travel times from maps and the actual size of objects from scale drawings.*
- *Use calculators to compare amounts proportionally.*

NCTM Standards 2000*
- *Solve problems involving scale factors, using ratio and proportion*
- *Develop, analyze, and explain methods for solving problems involving proportions, such as scaling and finding equivalent ratios*
- *Select appropriate methods and tools for computing with fractions and decimals from among mental computation, estimation, calculators or computers, and paper and pencil, depending on the situation, and apply the selected methods*
- *Recognize and apply geometric ideas and relationships in areas outside the mathematics classroom, such as art, science, and everyday life*

Math
Proportional reasoning
    rates
    scaling
Estimation

Integrated Processes
Observing
Collecting and organizing data
Predicting
Inferring
Interpreting data

Materials
Drawing compass
Ruler

Background Information
In the movie *Cast Away*, the main character, Chuck Noland, survives an airplane crash over the ocean and washes up on a deserted island. He has a monologue with his "companion," Wilson the volleyball. In this discussion he states they have traveled 11.5 hours from Memphis at 475 mph plus an extra 400 miles avoiding a storm. He draws a line of flight that shows their path flying over the northern edge of the Gulf of California.

Using the data given by Chuck, you can determine they traveled 5862.5 miles from Memphis (11.5 • 475 + 400 = 5862.5). Using a map and drawing the scaled range of flight with a compass centered in Memphis, Chuck can be placed along a large ring in the middle of the Pacific Ocean. Referring to the sketch in the movie, the flight flew over the northern edge of the Gulf of California. Using a straight edge and drawing a radius from Memphis to the ring that passes over the northern edge of the Gulf of California puts Chuck's position somewhere in the vast, empty ocean between the Hawaiian Islands and Guam. Drawing a circle with a scaled radius of 400 miles from Chuck's estimated position provides no evidence of a habitable atoll or island.

After Chuck escapes the island, he is told his island was 600 miles south of the Cook Islands. (We learn this information from his ex-girlfriend, Kelly Lovett, in a scene near the end of the movie.) This is near the circle of estimated distance, but not near the predicted location from the monologue. If the director had had the sketch show the flight path near the southern tip of the Baja Peninsula, it would have placed Chuck's prediction much closer his actual location.

This is at best an estimation. The Mercator projection map used in this activity has both distance and heading distortions. If a more accurate measure is desired, a globe can be substituted for the map. Slightly different positions will be identified, but the change is not significant.

Management

1. This activity is most motivating if students view the monologue on which it is based. It takes place about 74 minutes into the movie *Cast Away*. Obtain the video and have it set before class to show this segment.
2. Younger students can do this activity by marking a string along the scale and then swinging an arc with the string. Some students may choose to mark 11.5 sets of 500 miles to get an estimate.

Procedure

1. Play the video clip that shows the monologue of Chuck Noland's estimated position and discuss the *Key Question*.
2. Distribute the script of the scene and the map. Tell the students to use the data in the script to calculate the distance traveled before the crash.
3. Direct the students to scale this distance and then use a compass to draw an arc on the map with a radius of this scaled distance and the center at Memphis.
4. Ask the students to refer to the scene and draw a flight line with a straight edge that mimics Chuck's by passing over the northern edge of the Gulf of California. This line from Memphis should intersect the arc between the Hawaiian Islands and Guam on the map.
5. Have the students draw a circle centered at the predicted crash location that has a radius of a scaled 400 miles.
6. Have a class discussion on how accurate they believe the predicted location is, and what might have affected the accuracy.
7. Have students locate the correct position of 600 miles south of the Cook Islands and discuss what errors in Chuck's estimations resulted in the mistake.

Discussion

1. How can you determine how far the plane was from Memphis when it crashed?
   [D =rt, 5462.5 = 475 • 11.5]
2. How precise is your prediction? How much should you round it? [time could be ±15 min., the speed was not constant, 5500 would be reasonable]

3. How accurate was Chuck's comment that the search area was twice the size of Texas? [appears relatively accurate, Texas = 267,338 sq. mi.]
4. What misinformation in the movie caused your inaccurate prediction of location? [direction of travel]
5. How could the movie have been changed to more accurately portray what it claims happened? [draw the line near the southern Baja Peninsula]
6. What tricks or ideas about estimating can you learn from this example? [round, know assumptions, allow for precision of numbers, have a general feeling for the magnitude of numbers]

Extension

Have students draw flight lines between different destinations. Have them determine the distance, the length of the flight, and key landmarks along the flight.

* Reprinted with permission from *Principles and Standards for School Mathematics*, 2000 by the National Council of Teachers of Mathematics. All rights reserved.

WHERE IN THE WORLD IS CHUCK NOLAND?

Chuck Noland's Monologue to Wilson

We were in route from Memphis for eleven and a half hours, at four hundred seventy-five miles per hour, (draws a line from Memphis along the north edge of the Gulf of California and out into the ocean), so they think we were right here.

But we went out of radio contact and flew around that storm, for about what, an hour? So that's a distance of what? Four hundred miles.

Four hundred miles square. That's one hundred sixty thousand times pi, three point one four. Five hundred two, four … That's a search area of five hundred thousand square miles. That's twice the size of Texas.

They may never find us.

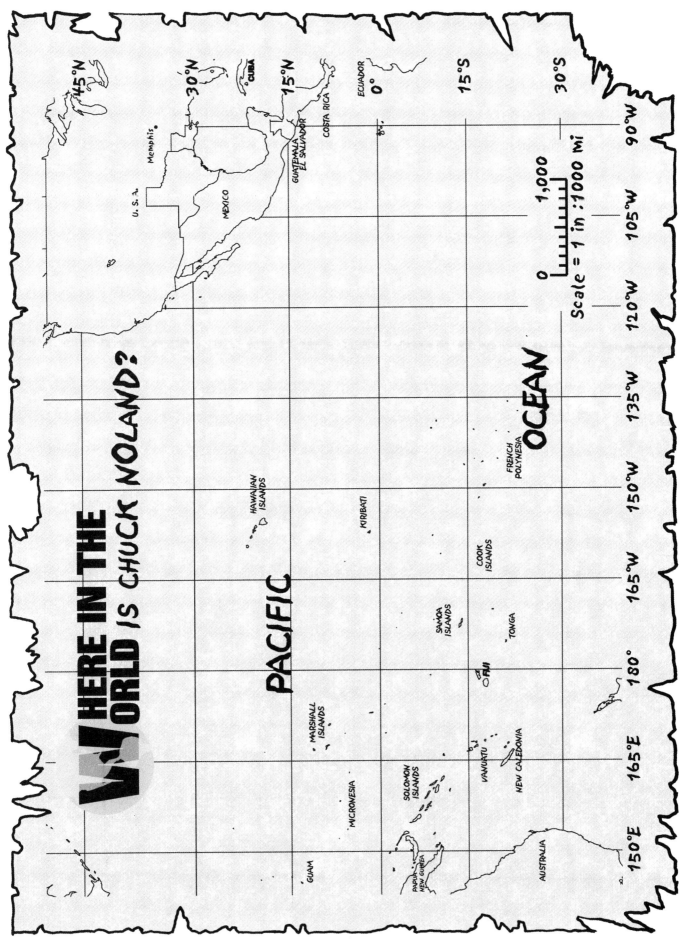

WHERE IN THE WORLD IS CHUCK NOLAND?

PACIFIC OCEAN

45°N 30°N 15°N ECUADOR 0° 15°S 30°S

90°W 105°W 120°W 135°W 150°W 165°W 180° 165°E 150°E

Memphis

U. S. A.

MEXICO

CUBA

GUATEMALA
EL SALVADOR

COSTA RICA

HAWAIIAN ISLANDS

KIRIBATI

FRENCH POLYNESIA

COOK ISLANDS

SAMOA ISLANDS

TONGA

FIJI

MARSHALL ISLANDS

VANUATU

NEW CALEDONIA

MICRONESIA

SOLOMON ISLANDS

GUAM

PAPUA NEW GUINEA

AUSTRALIA

0 1,000

Scale = 1 in :1000 mi

HOW MUCH ROPE?

Topic
Estimation

Key Question
How much rope can Chuck Noland make from the eight video tapes that washed up on the island?

Learning Goals
Students will:
1. make estimations using rates and proportions, and
2. make reasonable estimations of lengths.

Guiding Documents
Project 2061 Benchmarks
- *Mathematics is helpful in almost every kind of human endeavor–from laying bricks to prescribing medicine or drawing a face. In particular, mathematics has contributed to progress in science and technology for thousands of years and still continues to do so.*
- *Usually there is no one right way to solve a mathematical problem; different methods have different advantages and disadvantages.*
- *Decide what degree of precision is adequate and round off the result of calculator operations to enough significant figures to reasonably reflect those of the inputs.*

NCTM Standards 2000*
- *Select appropriate methods and tools for computing with fractions and decimals from among mental computation, estimation, calculators or computers, and paper and pencil, depending on the situation, and apply the selected methods*
- *Understand relationships among units and convert from one unit to another within the same system*
- *Solve simple problems involving rates and derived measurements for such attributes as velocity and density*

Math
Proportional reasoning
   rates
Estimation
Measurement
   linear

Integrated Processes
Observing
Predicting
Inferring
Interpreting data

Materials
Videotape cassette (VHS)
Yard or meter stick
100 ft. or 30 m tape
Philips screwdriver, optional

Background Information
In the movie *Cast Away*, the main character, Chuck Noland, survives an airplane crash over the ocean and washes up on a deserted island. After four years, he decides it is time to build a raft and escape from the island. He has several monologues with his "companion" Wilson, the volleyball, as he estimates how much rope he needs in order to lash a raft together and how much time he has to get ready. From the monologues and some simple calculations, it can be estimated he has enough time, but the materials for making rope are exhausted. Chuck decides to improvise and use videotape for rope. The obvious question arises, does he have enough videotape?

Earlier in the film Chuck unpacks the boxes that have washed ashore from the crash. One box contains eight cassettes of videotape. At standard play speed, the tape is used at the rate of about one and a third inches per second (approximately 3.3 cm/sec). If they are standard blank tapes with 60 minutes of recording time, the linear length of the tape can be calculated (1.33 inches/second x 60 seconds/minute = 80 inches/minute; 80 inches/minute x 60 minutes = 4800 inches /hour; 4800 inches ÷ 12 inches/foot = 400 feet or 133.3 yards. You might visualize this as one and a third football fields.)

The numbers in this calculation lend themselves to mental estimation. As students are challenged to think of being stranded without a calculator or pencil and paper and solving this problem, they might look at the problem differently. One possibility is to think of 60 seconds of 1 1/3 inches/second or 1 1/3 of 60. If you take a whole sixty and a third of 60, which is 20, and combine them, you get 80 inches a minute. Now 80 inches/minute times 60 minutes is relatively easy to do in your head as you multiply 8 times 6 and add two zeros, 4800 inches. Likewise, dividing by 12 inches per foot works out by dividing 48 hundred by 12 to get 4 hundred feet. Now I can take 300 of those feet along a 100-yard football field plus 100 more feet, that is a third of another football field.

This is at best an estimation of how much tape Chuck would get out of one cassette. Blank tapes are labeled with their recording times. They are normally sold in 30-minute increments. Commercially produced pre-recorded tapes vary in length. The length is sometimes recorded on the cassette, but more often you just get a title on the label. The videotapes in the movie might be any length. They might have been commercially sold blank cassettes that had meetings or conferences recorded on them. They might have been commercial tapes like training videos or short commercial promotions.

Management

1. This activity is most motivating if students view the monologue on which it is based. It takes place about 84 minutes into the movie *Cast Away*. Obtain the video and have it set before class to show this segment.
2. For this activity the students determine the length of tape on a specific cassette. Before class the teacher needs to find a cassette that is **no longer needed.** Determine its length from the label, or reset the counter on the VCR and play the tape to determine the recording time on the tape. Short promotional tapes are most reasonable in length. One hundred twenty minute blank tapes will be about 800 feet long.
3. To confirm the length of the tape, students need to remove the tape from the cassette. On blank tapes and longer commercially produced tapes, this is done by flipping open the front protective cover while you press the small button near the front of the left side. While the cover is open, push the point of a pen or pencil into the small quarter-inch hole near the back of the bottom of the cassette. A second person can now unwind the tape as needed. This method does not allow you to easily rewind the tape and use it again.

   A second method is to remove the five Philips screws from the bottom of the cassette. Open up the cassette and remove the two reels of tape. After measuring the length of the tape, it can be wound back on one or both of the reels.

Procedure

1. Play the video clip that shows the monologue of Chuck Noland's estimation of time until departure and amount of rope required. Students may refer to the script of his monologue.
2. Have the students discuss Chuck's estimation methods and talk about how they would go about estimating how long it would take him to make 475 feet of rope and how much time would remain for building, stocking, and launching the raft. Have students record their estimation methods.
3. Ask students how they might determine how long the tape in the sample videocassette is without pulling it out. Inform the students that one and

one-third inches of tape are used for each second of viewing, and give them the length of recording time that is provided on the sample tape.
4. Encourage the students to use mental estimation methods and estimate the length of the tape. Have them record their methods.
5. As a class, discuss the variety of tricks and methods used to estimate the length of the tape.
6. Have students make a comparison of their estimation of the tape's length to a distance of a location at school like the soccer field, baseball diamond, or basketball court.
7. Have the students confirm as a class the length of the tape by unwinding it and measuring it with a yardstick or 30 meter tape. It is effective to go out to the compared location at school. As a team unwinds the tape the length of the selected location, a second team can be measuring the length of the location.
8. When students are familiar with the length of the videotapes, have them discuss the *Key Question*.

Discussion

1. How can you estimate how much time is needed to make the 475 feet of rope? [475 ft. ÷ 15 ft./day = 31.666… days ≈ 32 days]
2. How can you estimate how much time Chuck has to build, stock, and launch the raft? [1.5 months ≈ 45 days, 45 days – 32 days for rope = 13 days, about 2 weeks]
3. What estimation tricks or methods did Chuck use? [rounding, 424 feet + 50 feet ≈ 475 feet; significant figures, 8 lashing times 24 feet per lashing ≈ 160 = 8 • 20]
4. How did you go about getting an estimation of how much videotape is in this cassette? (Answers will vary, see *Background Information*.)
5. What is something around the school that is as long as this videotape?
6. What are some situations for which an estimation is good enough, and other situations for which exact calculations are required?

Extension

Have students develop a project to estimate the duration of something using just a small sample for information. Examples might be, how long will it take you read a book if you have only read several pages? How many pennies are required to cover the classroom floor if you only have 50¢ worth? How long a line could you draw with a pencil? How big an area could you darken with a pencil?

* Reprinted with permission from *Principles and Standards for School Mathematics*, 2000 by the National Council of Teachers of Mathematics. All rights reserved.

HOW MUCH ROPE DO YOU GET FROM A VIDEOTAPE?

Chuck Noland's Monologue to Wilson

Twenty-two, 44 lashings. Forty-four lashings, so. We have to make rope again. Wilson, we're going to have to make a h--- of a lot of rope.

Eight lashings will be structural, so that's 24 a piece, that will be a hundred and, a hundred and sixty.

Here we are today. That gives us another month and a half until we're into March and April, which is our best chances for the high tides and the offshore breezes.

We need four hundred and twenty-four feet of good rope. Plus another 50 say for miscellaneous. That, round that off to 475 feet of good rope.

If we average 15 feet a day, plus we have to build it. We have to stock it. We have to launch it. That's gonna be tight. That is not much time, but we live and we die by time, don't we? Now let's not commit the sin of turning our back on time. I know, I know.

This is it. That's all that's left. I've checked over the whole island and that is all that's left. So we're going to be short. Short. Just have to make some more out of videotape. Yes. No, we have time. We do. We have time. Look the wind's still blowing in from the west.

HOW MUCH ROPE
DO YOU GET FROM A VIDEOTAPE?

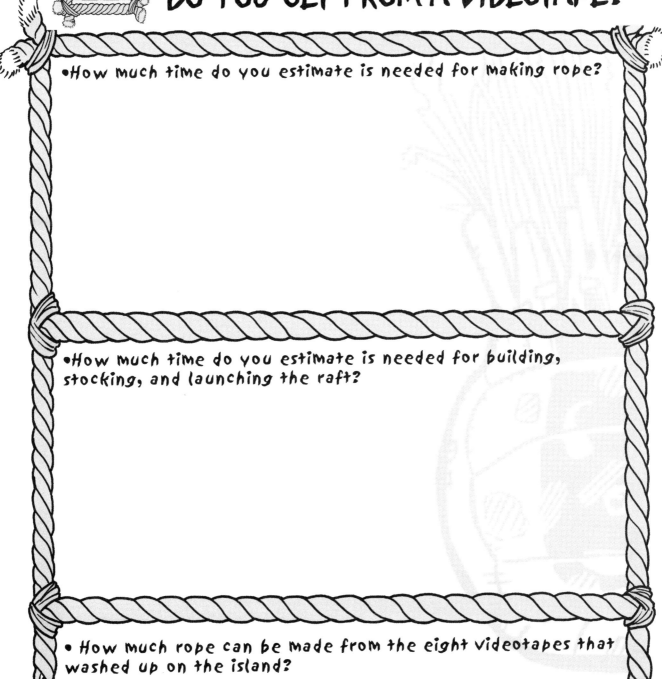

• How much time do you estimate is needed for making rope?

• How much time do you estimate is needed for building, stocking, and launching the raft?

• How much rope can be made from the eight videotapes that washed up on the island?

THE MATRIX

Matrix is a sci-fi action film in which the world is run by computers, and humans live virtual lives while providing energy for the computers. The hero, Neo, lives his virtual life as a young software engineer and part-time hacker who is singled out by some mysterious figures who want to introduce him to the secret of "the matrix." In the process of trying to confront "the matrix," Neo has to save the leader from a skyscraper. He and the heroine, Trinity, ride an elevator partway up the building where they stop the elevator and activate a bomb. They climb up and attach themselves to the elevator cable. As they release the elevator from the cable, it drops to explode while the counterweight lifts Neo and Trinity to the top of the building.

This elevator scene provides a context in which to study operations with integers and a use for inequalities. The clues provided within the clip allow students the chance to determine the mystery of the supposed height of this building.

Matrix received an R rating due to sci-fi violence and brief language. There is an edited version for television. The clip of the elevator scene, found in the second half of the movie, has no violence or objectionable language.

Elevator Integers

Topic
Integers

Key Question
If you know the motions of the counterweight or the elevator, how can you develop a system using numbers to keep track of the position of an elevator?

Learning Goals
Students will:
1. develop an understanding of the meaning and process of adding and subtracting integers, and
2. relate a visual model of adding and subtracting integers to solve problems.

Guiding Documents
Project 2061 Benchmarks
- *A number line can be extended on the other side of zero to represent negative numbers. Negative numbers allow subtraction of a bigger number from a smaller number to make sense, and are often used when something can be measured on either side of some reference point (time, ground level, temperature, budget).*
- *The operations + and - are inverses of each other— one undoes what the other does; likewise x and ÷.*

NCTM Standards 2000*
- *Develop meaning for integers and represent and compare quantities with them*
- *Understand the meaning and effects of arithmetic operations with fractions, decimals, and integers*
- *Understand and use the inverse relationships of addition and subtraction, multiplication and division, and squaring and finding square roots to simplify computations and solve problems*

Math
Number
   integers
   operations

Integrated Processes
Observing
Comparing and contrasting
Generalizing
Applying

Materials
Classroom model:
   80 inches of adding machine tape
   2 binder clips
   6' string
   meter stick
   masking tape
   2 index cards, 3" x 5"
   2 paper clips
   marker
   single hole punch

Student model:
   card stock
   11" of string
   2 pennies
   tape
   stapler
   single hole punch

Background Information
The addition and subtraction operations of integers pose difficulty for students because of a lack of understanding of the meaning of the operations or a misconception of the effect of the operations.

From experience most students have developed the concept that adding numbers means putting them together to get a larger number. Likewise they view subtraction as the operation on two numbers that results in a smaller number. When students are introduced to negative numbers, these misconceptions result in incorrect solutions and disequilibrium for the students. Models that help develop a fuller understanding of the operations are beneficial to introduce at this time.

A simple elevator with its counterweight provides a useful model. The "floors" of the elevator shaft form a number line. A number will tell a position or distance of motion along the number line. As the elevator is moved, it is quickly observed that the counterweight travels in the opposite direction of the elevator.

The situation of an elevator starting at the fourth floor and moving up seven floors is modeled with the number sentence 4 + 7 = 11. Likewise, an elevator starting at the ninth floor and going down three floors can be represented with the number sentence 9 - 3 = 6. These two situations model what is familiar numerically to students, but it is important that the students identify what the operations represent in the situations. An addition is an upward motion, and subtraction represents going down.

The situation of the elevator starting on the second floor and going down seven floors to stop five floors down in the basement (2 – 7 = -5) begins to trouble students. Although the number five in the solution is associated as the difference of seven and two, they would be more comfortable with positive five than with the negative five that results. When considering the situation of starting two floors down in the basements and going down three more results at being five floors down in the basement causes students little trouble. However, the number sentence -2 – 3 = -5 is disturbing because the solution is more familiar as an addition sum than a difference. Students will be much more successful with signed numbers if they are allowed to work with an elevator model to develop an understanding of why the solutions show up. Given many opportunities to translate elevator problems into numeric form will offer students the experiences necessary to make their own generalizations of how to solve numeric representations.

An elevator model also provides an understanding of the opposite effect of operating with negative numbers. The motion of the counterweight is opposite to the elevator's motion. The counterweight might be considered the negative elevator, or opposite elevator. So a motion of three (3) on the elevator would be a motion of three in the opposite direction (-3) on the counterweight. If the elevator is on the fourth floor (4) and the counterweight went up (+) three floors (-3) the elevator would end up on the first floor. Represented in a number sentence it would be: 4 + (-3) = 1. A counterweight going up means the elevator is going down. Adding an opposite is just like subtracting a positive. Working with the elevator model will reinforce the opposite nature of negative numbers and will provide a mental model to which students can refer when they forget the generalizations they have made.

Management
1. Before beginning the lesson, follow the instructions to construct a classroom elevator model. It is designed to hang in an open doorway or extend from a high cabinet.
   a. Cut an 80-inch (203-cm) strip of adding machine tape.
   b. Fold it in half five times to make 32 2.5-inch segments.
   c. With a marker, make a large G (for ground floor) or 0 on one of the middle segments.
   d. Draw a line with the marker at each fold between the segments and number the segments above 0 to 13, and below 0 to -13.

e. Punch a hole in the center of the line above 13 on the adding machine tape.
f. Use one binder clip to attach the numbered tape to the middle of the meter stick.
g. Move the numbered tape until the hole is just above the silver wire handles of the clip when the handles lie on either side of the tape.
h. Fold up bottom segment of the tape and attach the other clip so it is below the -13 segment.
i. Tape the meter stick across the door high enough to suspend the bottom clip and keep the adding machine tape taut.
j. Tie one end of a 6-foot long string to one paper clip and thread the other end through the hole within the top clip's handles.
k. Pull the string through until the paper clip is centered in the G segment (0) and tie the second paper clip to the end so both clips will hang centered in the G segment.
l. Fold the index cards in half so they are 1.5" x 5" and put them in the paper clips so they balance and extend on both sides of the numbered tape.

2. If you anticipate the students will need their own models, copy the *Student Elevator Model* onto card stock and have materials ready for students to construct them. The students can follow the instructions on the models.

3. The movie *Matrix* has a scene in which the hero, Neo, attaches himself to the cable of an elevator. He disengages the elevator and lets the counterweight lift him and the heroine to the top of a skyscraper. For motivation, this scene can be shown to students before the activity. Suggest that when they understand how elevators work, they should be able to determine the height of the building from the clues in the movie sequence. (Note: *Matrix* has an R rating for language and violence; however, this sequence has no violence or objectionable language. Check with your local district for viewing regulations. Follow any procedures necessary for gathering permission.)

4. This investigation might be spread over several days. Each student record sheet with its solution and discussion forms a daily lesson.

Procedure

1. If appropriate, show the movie sequence from *Matrix* and explain to the students that they need to understand how elevators work in order to determine the height of the building from clues in the movie.

2. Show the students the classroom model elevator that has been prepared beforehand and demonstrate that the motion of the elevator and counterweight are related.

3. Distribute the student page *Watching the Elevator*. Have the students consider each sentence of directions and use the model to get the correct solution. (The classroom model may be used by having one student move the elevator, or the students can be given materials and construct their own individual elevator model if that is appropriate.)

4. Have the students record the motion and positions of the elevator as a number sentence. Some students may require a little guidance to get started.

5. When students have completed the page, have them discuss how different words are translated into the number sentence and what strategies the students use to think about the solutions to the number sentence.

6. Distribute the student page *Watching the Counterweight*. Before beginning, make sure students have clearly identified that moving the counterweight up makes the elevator go down. Conversely moving the counterweight down makes the elevator go up. Suggest to students that they identify the counterweight as the "anti-elevator" or "opposite elevator" and that the subtraction (–) symbol also means opposite in math.

7. Using the classroom model, have a student move the elevator according to the first situation on the student page to determine the solution.

8. Have the students record the motion and positions of the elevator as a number sentence. You might suggest using (-3) to represent the motion of the "opposite elevator" three floors so the final sentence would be (2 + -3 = -1). Using the model, have the students complete the page.

9. Discuss with students how the number sentence represents the position and motions of the elevator and its counterweight.

10. Distribute the *Practice Pattern* page and have students complete it, referring to the model when necessary or to confirm their answers.

11. Discuss on patterns students notice in the solutions and strategies they could use to solve similar problems without the model.

Discussion

1. How are the motions of the elevator and counterweight related? [They move in opposite directions.]

2. How do each of the numbers and symbols represent what is happening with the elevator system? [First number: elevator's starting position; Operation: + = up, – = down; Second number: distance of motion (+ elevator, – counterweight), Solution: final elevator position]

3. What strategies can you use from the elevator model to solve number sentence? [Simplify the sentence using the opposite motion of the counterweight and elevator. The counterweight going up means the elevator goes down the same distance, and the counterweight going down means the elevator goes up the same distance. (To solve the number sentence, students will think about the general final position of the elevator and will manipulate the numbers to get it. Some students will state rules very close to those suggested by a typical text.)]

* Reprinted with permission from *Principles and Standards for School Mathematics*, 2000 by the National Council of Teachers of Mathematics. All rights reserved.

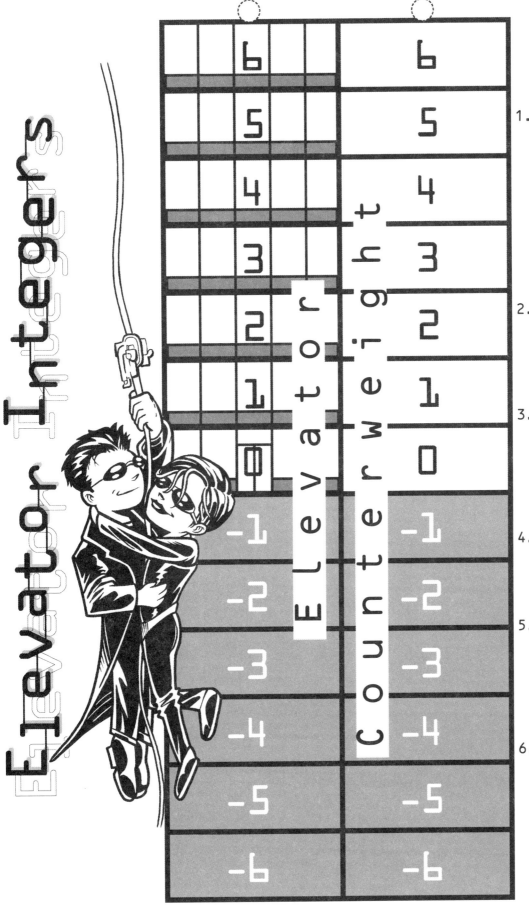

1. Fold the paper along the bold vertical lines to form a T. The two building faces should be back to back and facing outward.

2. Staple the top and the bottom of the building together.

3. Punch a hole where indicated. This is the center of the roof of the building.

4. Thread an 11-inch string through the hole.

5. Hold a penny in the center of the zero floor on both sides of the building.

6. Have a partner tape the string to the pennies so they hang in the zero position.

Elevator Integers

WATCHING THE ELEVATOR

Translate each sentence to the other form and use the elevator model to determine the solution.

	English Sentence	Number Sentence
1.	The elevator was on the second floor and went up three floors. Where is it?	
2.	The elevator was on the first floor and went up four floors. Where is it?	
3.	The elevator was on the sixth floor and went down four floors. Where is it?	
4.	The elevator was on the fourth floor and went down three floors. Where is it?	
5.	The elevator was on the second floor and went down three floors. Where is it?	
6.	The elevator was on the fifth floor and went down seven floors. Where is it?	
7.	The elevator was on the sixth floor and went down ten floors. Where is it?	
8.	The elevator was on the third floor and went down nine floors. Where is it?	
9.	The elevator was three floors down in the basement and went up four floors. Where is it?	
10.	The elevator was two floors down in the basement and went up seven floors. Where is it?	
11.	The elevator was five floors down in the basement and was needed on the fifth floor. How many floors up does it need to go?	
12.	The elevator was four floors down in the basement and was needed on the sixth floor. How many floors up does it need to go?	
13.	The elevator was on the second floor and was needed on the fourth floor down in the basement. How many floors does it need to move, and in what direction?	
14.	The elevator was on the sixth floor and was needed one floor down in the basement. How many floors up does it need to move, and in what direction?	
15.		$3 - 5 =$
16.		$-4 + 7 =$

Elevator Integers
WATCHING THE COUNTERWEIGHT

Translate each sentence to the other form and use the elevator model to determine the solution.

	English Sentence	Number Sentence
1.	The elevator was on the second floor and the counterweight went up three floors. Where is the elevator?	
2.	The elevator was on the fourth floor and the counterweight went up two floors. Where is the elevator?	
3.	The elevator was on the third floor and the counterweight went up one floor. Where is the elevator?	
4.	The elevator was two floors down in the basement and the counterweight went up four floors. Where is the elevator?	
5.	The elevator was five floors down in the basement and the counterweight went up one floor. Where is the elevator?	
6.	The elevator was one floor down in the basement and the counterweight went up four floors. Where is the elevator?	
7.	The elevator was on the fourth floor and the counterweight went down two floors. Where is the elevator?	
8.	The elevator was on the second floor and the counterweight went down four floors. Where is the elevator?	
9.	The elevator was on the third floor and the counterweight went down one floor. Where is the elevator?	
10.	The elevator was three floors down in the basement and the counterweight went down six floors. Where is the elevator?	
11.	The elevator was six floors down in the basement and the counterweight went down four floors. Where is the elevator?	
12.	The elevator was three floors down in the basement and the counterweight went down eight floors. Where is the elevator?	
13.		$-2 - -5 =$
14.		$1 + -3 =$
15.		$2 - -4 =$
16.		$-3 + -2 =$

Elevator Integers
PRACTICE PATTERNS

Use the elevator model to solve each problem. Then consider how you would get the solutions using arithmetic. Identify a method you can use to solve the number sentences without the model.

1.	4 + 2 =	-4 + 2 =	4 + -2 =	-4 + -2 =
2.	3 + 1 =	-3 + 1 =	3 + -1=	-3 + -1 =
3.	5 + 8 =	-5 + 8 =	5 + -8 =	-5 + -8 =
4.	8 + 3 =	-8 + 3 =	8 + -3 =	-8 + -3 =
5.	6 + 6 =	-6 + 6 =	6 + -6 =	-6 + -6 =
6.	4 + 7 =	-4 + 7 =	4 + -7 =	-4 + -7 =

7.	3 − 2 =	-3 − 2 =	3 − -2 =	-3 − -2 =
8.	5 − 4 =	-5 − 4 =	5 − -4 =	-5 − -4 =
9.	4 − 7 =	-4 − 7 =	4 − -7 =	-4 − -7 =
10.	6 − 10 =	-6 − 10 =	6 − -10 =	-6 − -10 =
11.	5 − 5 =	-5 − 5 =	5 − -5 =	-5 − -5 =
12.	7 − 5 =	-7 − 5 =	7 − -5 =	-7 − -5 =

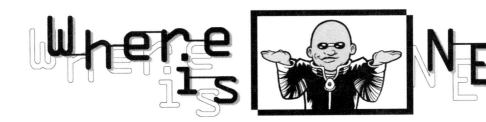

Topics
Integers
Indirect variation

Key Question
If you know the position of the elevator, how can you determine where Neo is?

Learning Goals
Students will:
1. generalize a pattern using variables and an equation,
2. graph a pattern and interpret the meaning the graph represents, and
3. recognize that the motion of an elevator and counterweight vary oppositely from each other (indirect variation).

Guiding Documents
Project 2061 Benchmarks
- *A number line can be extended on the other side of zero to represent negative numbers. Negative numbers allow subtraction of a bigger number from a smaller number to make sense, and are often used when something can be measured on either side of some reference point (time, ground level, temperature, budget).*
- *The operations + and – are inverses of each other—one undoes what the other does; likewise x and ÷.*
- *Mathematicians often represent things with abstract ideas, such as numbers or perfectly straight lines, and then work with those ideas alone. The "things" from which they abstract can be ideas themselves (for example, a proposition about "all equal-sided triangles" or "all odd numbers").*

NCTM Standards 2000*
- *Develop meaning for integers and represent and compare quantities with them*
- *Understand and use the inverse relationships of addition and subtraction, multiplication and division, and squaring and finding square roots to simplify computations and solve problems*
- *Represent, analyze, and generalize a variety of patterns with tables, graphs, words, and, when possible, symbolic rules*

Math
Number
  integers
Algebraic thinking
  variables
  equations
  graphs
  indirect variation

Integrated Processes
Observing
Comparing and contrasting
Generalizing
Applying

Materials
Classroom model:
  80 inches of adding machine tape
  2 binder clips
  6' string
  meter stick
  masking tape
  2 index cards, 3" x 5"
  2 paper clips
  marker
  single hole punch

Background Information
The movie *Matrix* has a scene in which the hero, Neo, attaches himself to the cable of an elevator. He disengages the elevator and lets the counterweight lift him and the heroine to the top of a skyscraper. This scene can be shown to students before the activity. Suggest that when they understand how elevators work, they should be able to determine the height of the building from the clues in the movie sequence.

For this activity a piece of masking tape will represent Neo. It will be stuck to the counterweight string where students cannot see it. The object is to determine where Neo is when the position of the elevator is known. For set-up, place the classroom model elevator at 0 or G. Stick the tape (Neo) to the string of the counterweight at the fourth floor. Ask a student to come up and move the elevator to a new position and inform the class where Neo is. After two or three examples, ask the students to predict where Neo is.

In the example, if the elevator were moved up to floor 5, Neo would go down 5 to -1. A few examples might include: elevator = -7, Neo = 11; elevator = 9, Neo = -5. The sum of the elevator's position and Neo's position for the three samples is 4. This pattern could be written in one of three forms: $E + N = 4$, $4 - N = E$, $4 - E = N$. The graph of Neo's position to the elevator's position is a straight line with a y-intercept and x-intercept of 4. This graph shows the indirect variation that as the elevator goes up, Neo goes down.

Moving Neo to a new position on the string will provide similar results. The sum of the elevator's position and Neo's position will be a constant. The relationship could be written in standard form as $E + N$ = Constant. The graph will be a line with a slope of -1, with y and x intercepts equal to the constant.

The actual building in Sydney, Australia where *Matrix* was filmed has 82 floors. This is seen as Neo enters the express elevator that services the floors 40 to 82. Neo, however, stops the elevator at the 41st floor and then proceeds up to where the counterweight passes him. This suggests that they did not stop in the middle of the building. Since the counterweight will fall more than 41 floors, Neo will rise more than 41 floors. One can count the banks of lights, one for each floor, as Neo rises. Because of changes in camera angles, the count is an estimate of 8 floors from the 41st to where the counterweight and Neo meet. They meet at about 49, meaning they are halfway up the building thereby suggesting that the building is about 98 floors high. Since this is an estimate, we might be better served by stating the range we are confident in as $41 < h < 98$.

Some students may suggest counting the bank of buttons inside the elevator. Not all the floors are represented in this bank because it is an express elevator and has no buttons between the ground and floor 40.

Before students get too excited about finding a mistake in the movie, remind them that the goal of the director was not as much accuracy as it was to show movement. That is why the counterweight is shown going by.

Management

1. The scene from the movie *Matrix* described in the *Background Information* provides the context for this activity. Before the activity find the sequence to show in class. (Note: *Matrix* has an R rating for language and violence, however this sequence has no violence or objectionable language. Check with your local district for viewing regulations. Follow any procedures necessary for gathering permission.)

2. Before beginning the lesson, follow the instructions below and construct a classroom elevator model. It is designed to hang in an open doorway or extend from a high cabinet.

   a. Cut an 80-inch (203-cm) strip of adding machine tape.

   b. Fold it in half five times to make 32 2.5-inch segments.

   c. With a marker, make a large G (for ground floor) or 0 on one of the middle segments.

   d. Draw a line with the marker at each fold between the segments and number the segments above 0 to 13, and below 0 to -13.

   e. Punch a hole in the center of the line above 13 on the adding machine tape.

   f. Use one binder clip to attach the numbered tape to the middle of the meter stick.

   g. Move the numbered tape until the hole is just above the silver wire handles of the clip when the handles lie on either side of the tape.

   h. Fold up bottom segment of the tape and attach the other clip so it is below the -13 segment.

   i. Tape the meter stick across the door high enough to suspend the bottom clip and keep the tape taut.

   j. Tie one end of a 6-foot long string to one paper clip and thread the other end through the hole within the top clip's handles.

   k. Pull the string through until the paper clip is centered in the G segment (0) and tie the second paper clip to the end so both clips will hang centered in the G segment.

   l. Fold the index cards in half so they are 1.5" x 5" and put them in the paper clips so they balance. The card on the front representing the elevator should extend on both sides of the numbered tape. The card in the back of the number line, representing the counterweight, should be placed vertically so its position is not obvious from the front.

Procedure

1. If appropriate, show the movie sequence from *Matrix* and explain to the students that they need to understand how elevators work in order to determine the height of the building from clues in the movie. If the movie sequence is not shown, explain the concept.
2. Show the students the elevator model that has been prepared beforehand and explain that their job is to determine at what floor Neo is.
3. Have a student volunteer move the elevator to a new position.
4. Inform the students of Neo's position while they record his position and the elevator's.
5. Allow the student to gather two or three data points before asking them to predict Neo's position.
6. Have the students discuss how they determined Neo's position and represent their solution method in an equation.
7. Ask the students to graph the data and discuss the relationship of the equation, the graph, and elevator system.
8. Continue to move Neo and have students record and graph data.
9. After Neo has been in several positions on the string, have the students generalize the relationship of the equations, graphs, and motion.

Discussion

1. What do you notice about the pairs of positions for the elevator and Neo? [They always add up to the same number, a constant.]
2. How do you predict Neo's position? [You subtract the elevator's position from the constant.]
3. How could you describe your method of prediction as an equation? [constant(C) – elevator(E) = Neo(N)]
4. What are some equivalent ways to write this equation? [E + N = C, C – N = E]
5. Describe the graph. [straight line, slope = -1, y and x intercepts = constant]
6. How does the graph interpret the motion of the elevator and Neo? [As the elevator goes higher, Neo goes lower.]
7. How are the data similar no matter where Neo is on the cable? [positions of elevator and Neo equal a constant, graph is a line with a slope of -1 going through intercepts equal to the constant]

Extension

View the elevator sequence from *Matrix* again to determine the height of the building.

* Reprinted with permission from *Principles and Standards for School Mathematics*, 2000 by the National Council of Teachers of Mathematics. All rights reserved.

WHERE IS  NEO?

Elevator	Neo		Elevator	Neo		Elevator	Neo		Elevator	Neo

Neo's Position

Elevator's Position

-12 -10 -8 -6 -4 -2 0 2 4 6 8 10 12

12 10 8 6 4 2 0 -2 -4 -6 -8 -10 -12

COOL RUNNINGS

The movie, Cool Runnings, is loosely based on the improbable but true story of Jamaica's first bobsled team. This team, comprised of a helicopter pilot, a reggae singer, and a sprinter, took part in the 1988 Winter Olympics in Calgary, Alberta, Canada. The movie catches the excitement and thrill of the bobsled event while capturing the desire of individuals to be recognized for their hard work.

As the movie shows, the Jamaican team did not complete its third run and as a result was disqualified from the event. Records of their first two runs at the Olympics allow students to consider the accuracy of the film. By looking at these records, students can begin to see the speeds and acceleration of the bobsled and consider if the Jamaicans were truly a competitive team or were rightly the joke of the Olympics.

JAMAICAN BOBSLED TEAM
PART ONE

Topic
Statistics

Key Question
How did the times of the Jamaican Bobsled team compare to the Olympic field? Were they competitive?

Learning Goals
Students will:
1. Use data from the 1988 Calgary Olympics to determine measures of central tendency and spread, and
2. Construct a box plot to compare the times of competitors.

Guiding Documents
Project 2061 Benchmarks
- *In the absence of retarding forces such as friction, an object will keep its direction of motion and its speed. Whenever an object is seen to speed up, slow down, or change direction, it can be assumed that an unbalanced force is acting on it.*
- *The mean, median, and mode tell different things about the middle of a data set.*
- *Comparison of data from two groups should involve comparing both their middles and the spreads around them.*

NRC Standards
- *Think critically and logically to make the relationships between evidence and explanations.*
- *If more than one force acts on an object along a straight line, then the forces will reinforce or cancel one another, depending on their direction and magnitude. Unbalanced forces will cause changes in the speed or direction of an object's motion.*
- *Mathematics is important in all aspects of scientific inquiry.*

NCTM Standards 2000*
- *Select, create, and use appropriate graphical representations of data, including histograms, box plots, and scatterplots*

- *Find, use, and interpret measures of center and spread, including mean and interquartile range*
- *Discuss and understand the correspondence between data sets and their graphical representations, especially histograms, stem-and-leaf plots, box plots, and scatterplots*

Math
Statistics
  measures of central tendency and spread
  box plots

Science
Physical science
  balanced forces

Integrated Processes
Observing
Comparing and contrasting
Collecting and organizing data
Predicting
Inferring
Interpreting data

Materials
Student pages

Background Information
    In the 1988 Winter Olympics in Calgary, the Jamaican Bobsled team made its debut. The movie *Cool Runnings* presents the general attitude towards this team from the tropics. Although many looked at the team as a joke, the team came through with final times typical for the range of Olympic results listed below.

<u>1988 Olympic Bobsled Times</u>
Lower Extreme: 55.88 sec.
Lower Quartile: 57.10 sec.
          Mode: 56.41 sec.
          Mean: 57.6057 sec.
        Median: 57.62
Upper Quartile: 58.135
Upper Extreme: 59.67

Jamaican Bobsled Team: 58.04, 59.37

The Olympic Bobsled event is the combined time of four runs down the track. Two runs are made on two different days. To get a final score, all four runs must be completed. The Jamaican team crashed on their fourth and final run so they did not receive an official time. The times are the unofficial times of the first two runs on February 27, 1988.

The finishes of bobsled events are decided by hundredths of a second. The difference of the best and worst times of any event at this level is less than four seconds.

The differences in the ranges of times can be understood by considering the physics of bobsledding. The fastest speed at which a gravity-powered object can move (terminal velocity) is determined by the balance of forces. When the resistive forces of friction and air resistance equal the gravitational force, an object will continue at a constant velocity; it does not accelerate. If two objects such as bobsleds have the same resistive forces (four runners and the same frontal area for air resistance), but different gravitational forces (weight of the sled and riders), they will accelerate differently. The sled with the greater mass will overcome the resistive forces more easily, will have a higher terminal velocity, and will accelerate more rapidly.

The rules in the bobsled events are an attempt to maintain equal resistive forces between sleds. In order to win, the team must focus on the critical components of the initial push start, getting in an aerodynamic position, and maintaining a line of least resistance. The difference in Olympic times is a result of these three components.

How would you decide what is the "normal" or "typical" time for a bobsled event? To get at this idea, you might determine the time of a sled in the "middle" of all the times by looking at measures of central tendency: mean (arithmetic average), median (middle of an ordered list), and mode (most often occurring value). These measures will give you only one number. How would you determine the range of very good times? Now you are grappling with the idea of distribution. To describe what to expect in a group, you need to have an idea of central tendency and distribution for the group.

Distribution can be studied by constructing an ordered list of the data and dividing it into four even groups. The value of the position that divides the data into two even groups is the median. The positions that divide the lower quarter and the upper quarter from the middle half are called the lower quartile and upper quartile respectively.

A box and whisker plot is one way to display distribution. It can be made on graph paper by making a horizontal number line scaled for the range of the data. On a line several squares above the number line, students can place dots to represent the median, the two quartiles, and the slowest and fastest times (extremes). A box can then be drawn around the middle half of the data leaving the quartiles at both ends. A vertical line is drawn in the box at the median position. Lines or "whiskers" are extended from each end of the box to the extremes.

Box and whisker plots can cause some difficulty for students. The students often expect the quarters to be the same size since the data are divided into four even groups. They misinterpret the quarters as the size of each group rather than the range in the value of data for each quarter of the data. Students will often need to consider the meaning of box and whisker plots several times before they can clearly interpret them.

Management

1. The data are placed in a sorted list to help students find the median and quartiles. They might find it helpful to circle or mark each measure in a list.
2. Listings of the official and unofficial times are included should a teacher choose to provide the students with the primary source. This provides the students the opportunity to work with a large source of data.

Procedure

1. Distribute the data sheet to the students and discuss the *Key Question* with the class.
2. Using the data sheet, have the students calculate and record the mean, median, and mode for the 1988 four-man bobsled event.
3. Hold a class discussion as to how the measures of central tendency help answer the *Key Question*.

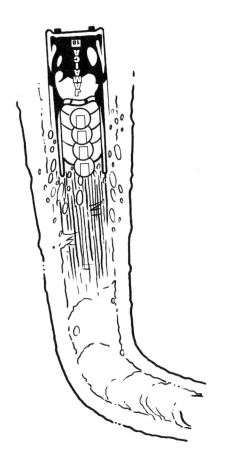

4. Have students identify and record the median, quartiles, and extremes for the event.
5. Direct the students to construct a box and whisker plot for the event.
6. Referring to their box plot, have students draw conclusions about the 1988 Olympic bobsled event and how the Jamaican team would have finished.

Discussion

1. What measure of central tendency does not help you much in making a generalization about the Olympics? [mode, not related to distribution]
2. Why is the mean not very useful in drawing conclusions for an Olympic event? [Rank is the only thing that matters in the Olympic decision.]
3. What generalizations can you make for the bobsled events by looking at your box and whisker plot? (Answers will vary.)
4. What might explain the differences in the ranges of times? (Refer to *Background Information*.)
5. How competitive was the Jamaican bobsled team in the 1988 Winter Olympics?

Extensions

1. Use the data from the 2002 Salt Lake Olympics to see how the Jamaican two-man Bobsled team did. The times are listed on the Internet and the International Bobsleigh Federation site:

   http://www.bobsleigh.com/index.htm

2. The Jamaican Bobsled Federation also has a site where more recent information is available:

   http://www.jamaicanbobsled.com

JAMAICAN BOBSLED TEAM
PART ONE

Official 1988 Olympic 4-Man Bobsled Times

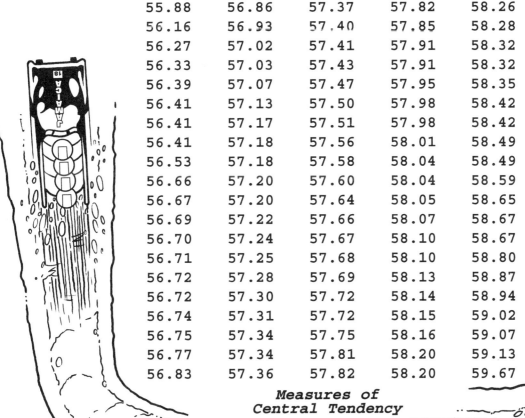

55.88	56.86	57.37	57.82	58.26
56.16	56.93	57.40	57.85	58.28
56.27	57.02	57.41	57.91	58.32
56.33	57.03	57.43	57.91	58.32
56.39	57.07	57.47	57.95	58.35
56.41	57.13	57.50	57.98	58.42
56.41	57.17	57.51	57.98	58.42
56.41	57.18	57.56	58.01	58.49
56.53	57.18	57.58	58.04	58.49
56.66	57.20	57.60	58.04	58.59
56.67	57.20	57.64	58.05	58.65
56.69	57.22	57.66	58.07	58.67
56.70	57.24	57.67	58.10	58.67
56.71	57.25	57.68	58.10	58.80
56.72	57.28	57.69	58.13	58.87
56.72	57.30	57.72	58.14	58.94
56.74	57.31	57.72	58.15	59.02
56.75	57.34	57.75	58.16	59.07
56.77	57.34	57.81	58.20	59.13
56.83	57.36	57.82	58.20	59.67

Measures of Central Tendency

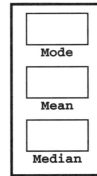

Mode

Mean

Median

Lower Extreme	Lower Quartile	Upper Quartile	Upper Extreme

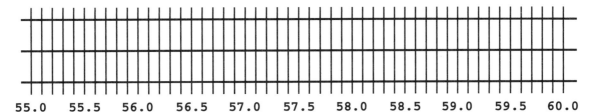

55.0   55.5   56.0   56.5   57.0   57.5   58.0   58.5   59.0   59.5   60.0

Jamacian Bobsled Team Times: 58.04   59.37

37

JAMAICAN BOBSLED TEAM
PART TWO

How does the speed of a bobsled change in a race?

Topic
Algebraic Thinking

Key Question
How did the speed of the Jamaican's bobsled change in the race?

Learning Goals
Students will:
1. use data from the 1988 Calgary Olympic Bobsled Race results to determine the average speeds on different segments of the race,
2. display the data on a distance versus time graph, and
3. learn that a steeper slope on the graph represents a greater speed or rate of change.

Guiding Documents
Project 2061 Benchmarks
- *In the absence of retarding forces such as friction, an object will keep its direction of motion and its speed. Whenever an object is seen to speed up, slow down, or change direction, it can be assumed that an unbalanced force is acting on it.*
- *Mathematical statements can be used to describe how one quantity changes when another changes. Rates of change can be computed from magnitudes and vice versa.*
- *Graphs can show a variety of possible relationships between two variables. As one variable increases uniformly, the other may do one of the following: always keep the same proportion to the first, increase or decrease steadily, increase or decrease faster and faster, get closer and closer to some limiting value, reach some intermediate maximum or minimum, alternately increase and decrease indefinitely, increase and decrease in steps, or do something different from any of these.*

NRC Standards
- *If more than one force acts on an object along a straight line, then the forces will reinforce or cancel one another, depending on their direction and magnitude. Unbalanced forces will cause changes in the speed or direction of an object's motion.*
- *Mathematics is important in all aspects of scientific inquiry.*

NCTM Standards 2000*
- *Explore relationships between symbolic expressions and graphs of lines, paying particular attention to the meaning of intercept and slope*
- *Use graphs to analyze the nature of changes in quantities in linear relationships*
- *Solve simple problems involving rates and derived measurements for such attributes as velocity and density*

Math
Algebra
  rates
Graphing

Science
Physical science
  speed
  forces

Integrated Processes
Observing
Comparing and contrasting
Predicting
Collecting and organizing data
Interpreting data
Inferring

Materials
Student pages

Background Information
In the 1988 Winter Olympics held in Calgary the Jamaican Bobsled team made its debut. The movie *Cool Runnings* presents the general attitude towards this team from the tropics. Although many looked at the team as a joke, the team came through with one race being near the top two-thirds of the field. Below are the splits for the Jamaican team's fastest race.

Time (sec)	Distance (m)	(ft.)
5.34	50	164
26.95	400	1312
36.19	800	2624
48.44	1200	3936
58.04	1475	4838

As in all races, the bobsled is determined by elapsed time. The finish of a bobsled event is often decided by hundredths of a second. It is critical for racers to determine when and where along the track they are losing time. For this reason, the track has been split into segments where time is measured with electric switches.

Speed is not used to determine who wins the race, time is. Although a sled may have been fastest at a certain spot, it may have been going slower at other times. Races are about average speed, not instantaneous speed.

Split times allow racers to look at different segments of the race and determine in which segments they could improve their average speed. The average speed for each segment can be determined by dividing the length of the segment by the time it took to travel that distance.

The concept of average speed begins to make sense as students compare numeric and graphic representations to their experience. By watching a video clip of a bobsled run, students understand the acceleration of the sled. They can relate it to their experiences on roller coasters or sledding. By taking the length of the track and dividing it by run time, the average speed of the sled can be determined. As students plot the starting and ending times and distances, they are quick to disagree with the representation. The straight line that is plotted shows the sled was traveling at a constant speed. As they graph and calculate the split speeds, they will agree that the average speed increases as the sled goes down the hill. As they calculate and record the average speed for each split, they will quickly see the relationship of the slope of the graph to the speed of the sled.

Management

1. Before the activity, you may wish to show segments of the video *Cool Runnings* to help students become familiar with a bobsled race.

2. Feet measurements have been included along with the metric measures used in the Olympics. Most students have no "feel" for meters per second and the teacher may have students calculate speeds in feet per second and convert it to the more familiar miles per hour.

Procedure

1. Show some video clips of a bobsled run (see *Management*) and discuss *the Key Question* with the class.
2. Distribute the student pages and have the students calculate and graph the average speed for the whole race.
3. Discuss with the class what the graph communicates and have them consider if it matches their experience and observations.
4. Have the students calculate and graph the speed of each split of the race.
5. Discuss with the class how the graphs differ and which is a better representation of the event.

Discussion

1. From looking at the videos or from your own experience, how do you expect the speed on the bobsled to change as it is ridden down the track? [faster and faster]
2. What does the straight line of the average speed graph say about the speed of the sled? [It does not change.]
3. How do the split speeds change through the race? [faster for the first two or three segments, then little change]
4. How do the changes in split speeds show up on the graph? [getting faster = steeper slope, little change = no change in slope]

Extensions

1. Have the students draw a graph for the fastest team in the 1988 Olympics and compare them to the Jamaican team.
2. Check the international bobsled organization website for recent results from Calgary and see if the field of bobsled teams is improving.

International bobsled organization
http://www.bobsleigh.com/index.htm

Calgary 88—Four Man Bobsled Official Results

Pos.	No.	Name	Nat.	50M	400M	800M	1200M	FINAL	POS	TIME DIFF TMPS	AVG KM/H MOY
1	21	Fasser E.	SUI-I	5.23	26.24	35.35	47.41	56.83	7	0.67	93.4
		Meier K.		5.22	26.44	35.63	47.82	57.37	3	0.09	92.5
		Faessler M.		5.21	26.00	34.95	46.73	55.88	1	0.00	95.0
		Stocker W.		5.17	26.06	35.26	47.68	57.43	4	0.23	92.4
							Total	3:47.51			
2	3	Hoppe W.	GDR-I	5.26	26.17	35.13	46.94	56.16	1	0.00	94.5
		Schauerhammer D.		5.24	26.46	35.62	47.78	57.31	2	0.03	92.6
		Musiol B.		5.26	26.22	35.29	47.35	56.77	13	0.89	93.5
		Voge I.		5.23	26.11	35.27	47.65	57.34	3	0.14	92.6
							Total	3:47.58			
3	14	Kipours I.	URS-II	5.28	26.19	35.27	47.32	56.72	4	0.56	93.6
		Ossis G.		5.19	26.41	35.59	47.78	57.28	1	0.00	92.7
		Tone I.		5.20	26.10	35.12	47.09	56.41	4	0.53	94.1
		Kozlov V.		5.15	26.15	35.43	48.00	57.85	7	0.65	91.7
							Total	3:48.26			
4	11	Rushlaw B.	USA-I	5.26	26.31	35.38	47.36	56.72	4	0.56	93.6
		Hoye H.		5.25	26.59	35.81	48.09	57.67	8	0.39	92.0
		Wasko M.		5.28	26.24	35.32	47.32	56.69	9	0.81	93.6
		White W.		5.27	26.22	35.36	47.61	57.20	1	0.00	92.8
							Total	3:48.28			
5	12	Poikans M.	URS-I	5.20	26.28	35.36	47.39	56.75	6	0.59	93.5
		Kliavinch O.		5.15	26.47	35.73	48.04	57.66	7	0.38	92.0
		Berzoups I.		5.18	26.10	35.19	47.27	56.70	10	0.82	93.6
		Iaoudzems I.		5.19	26.20	35.36	47.64	57.24	2	0.04	92.7
							Total	3:48.35			
6	22	Kienast P.	AUT-I	5.19	26.34	35.47	47.58	57.07	11	0.91	93.0
		Siegl F.		5.24	26.42	35.63	47.84	57.40	4	0.12	92.5
		Mark C.		5.21	26.07	35.07	46.96	56.27	2	0.39	94.3
		Teigl K.		5.18	26.10	35.34	47.98	57.91	8	0.71	91.6
							Total	3:48.65			
7	17	Appelt I.	AUT-II	5.26	26.26	35.37	47.46	56.93	8	0.77	93.2
		Muigg J.		5.28	26.50	35.71	47.91	57.51	5	0.23	92.3
		Redl G.		5.24	26.06	35.09	47.05	56.41	4	0.53	94.1
		Winklet H.		5.26	26.21	35.49	48.14	58.10	13	0.90	91.3
							Total	3:48.95			
8	24	Richter D.	GDR-II	5.19	26.24	35.43	47.62	57.18	13	1.02	92.8
		Ferl B.		5.21	26.39	35.62	47.94	57.60	6	0.32	92.1
		Jahn L.		5.26	26.13	35.11	47.01	56.33	3	0.45	94.2
		Szelig A		5.27	26.25	35.50	48.08	57.95	9	0.75	91.6
							Total	3:49.06			
										TIME	AVG

Pos.	No.	Name	Nat.	50M	400M	800M	1200M	FINAL	POS	DIFF TMPS	KM/H MOY
						TIMES/TEMPS					
9	4	Hilterbrand H.	SUI-II	5.25	26.25	35.25	47.13	56.39	2	0.23	94.1
		Fehlmann U.		5.25	26.65	35.91	48.23	57.91	12	0.63	91.6
		Fassbind E.		5.27	26.33	35.47	47.61	57.13	15	1.25	92.9
		Kiser A.		5.30	26.41	35.62	48.03	57.82	6	0.62	91.8
							Total	3:49.25			
10	20	Wolf A.	ITA-I	5.21	26.34	35.50	47.681	57.20	15	1.04	92.8
		Gesuito P		5.22	26.48	35.74	48.06	57.72	9	0.44	91.9
		Beikircher G.		5.22	26.23	35.29	47.24	56.53	7	0.65	93.9
		Ticci S.		5.22	26.21	35.48	48.10	58.01	10	0.81	91.5
							Total	3:49.46			
11	15	Fischer A.	FRG-II	5.28	26.40	35.51	47.59	57.02	9	0.86	93.1
		Niessner F.		5.27	26.56	35.79	48.12	57.75	10	0.47	91.9
		Eisenreich U.		5.25	26.21	35.27	47.30	56.71	11	0.83	93.6
		Langen C.		5.23	26.32	35.61	48.19	58.07	12	0.87	91.4
							Total	3:49.55			
12	10	Tout M.	GBR-I	5.28	26.39	35.52	47.70	57.22	16	1.06	92.7
		Armstrong D.		5.31	26.71	36.01	48.49	58.26	17	0.98	91.1
		Paul L.		5.28	26.21	35.31	47.41	56.86	14	0.98	93.3
		Richards A.		5.29	26.30	35.50	47.85	57.56	5	0.36	92.2
							Total	3:49.90			
13	16	Haydenluck G	CAN-II	5.21	26.44	35.60	47.71	57.18	13	1.02	92.8
		Langford C.		5.26	26.64	35.88	48.15	57.82	11	0.54	91.8
		Tyler K.		5.21	26.20	35.27	47.27	56.67	8	0.79	93.7
		Guss L.		5.18	26.23	35.56	48.30	58.32	17	1.12	91.0
							Total	3:49.99			
14	26	Sperr M.	FRG-I	5.24	26.48	35.70	47.96	57.58	19	1.42	92.2
		Hampel O.		5.28	26.64	35.94	48.33	58.04	13	0.76	91.4
		Cruciger F.		5.24	26.14	35.17	47.09	56.41	4	0.53	94.1
		Mueller R.		5.24	26.35	35.62	48.22	58.14	15	0.94	91.3
							Total	3:50.17			
15	2	Lori C.	CAN-I	5.23	26.24	35.27	47.26	56.66	3	0.50	93.7
		Leblanc K.		5.28	26.63	35.92	48.32	58.05	14	0.77	91.4
		Swim A.		5.24	26.25	35.43	47.70	57.34	18	1.46	92.6
		Dell H.		5.30	26.36	35.68	48.36	58.32	17	1.12	91.0
							Total	3:50.37			
16	9	Roy M.	USA-II	5.35	26.59	35.70	47.76	57.17	12	1.01	92.8
		Pladel S.		5.33	26.91	36.28	48.74	58.49	19	1.21	90.7
		Herberich J.		5.37	26.54	35.74	47.94	57.47	19	1.59	92.3
		Shimer B.		5.39	26.59	35.85	48.32	58.10	13	0.90	91.3
							Total	3:51.23			
17	23	De La Hunty T.	GBR-II	5.22	26.56	35.83	48.16	57.81	22	1.65	91.8
		Rattigan C.		5.26	26.68	36.00	48.42	58.13	15	0.85	91.3
		Robertson G.		5.26	26.28	35.35	47/35	56/74	12	0.86	93.5
		Leonce A.		5.30	26.61	35.98	48.67	58.59	20	1.39	90.6
							Total	3:51.27			
								TIME			AVG

Pos.	No.	Name	Nat.	50M	400M	800M	1200M	FINAL	POS	DIFF TMPS	KM/H MOY
							TIMES/TEMPS				
18	7	Sakai T.	Jpn-I	5.43	26.83	35.95	47.99	57.36	17	1.20	92.5
		Wakita T.		5.41	26.91	36.19	48.51	58.15	16	0.87	91.3
		Yaku Y.		5.38	26.68	35.85	48.11	57.68	22	1.80	92.0
		Takewaki N.		5.40	26.65	35.90	48.39	58.16	16	0.96	91.2
							Total	3:51.35			
19	13	D Amico R.	ITA-II	5.33	26.61	35.81	48.03	57.69	20	1.53	92.0
		Rottensteiner T.		5.35	26.85	36.20	48.78	58.65	21	1.37	90.5
		Scaramuzza P.		5.35	26.49	35.66	47.90	57.50	20	1.62	92.3
		Meneghin A.		5.38	26.58	35.81	48.24	58.04	11	0.84	91.4
							Total	3:51.88			
20	6	Nagy Lakatos C	ROM-I	5.26	26.47	35.63	47.82	57.41	18	1.25	92.4
		Grigore A		5.24	26.74	36.08	48.61	58.49	19	1.21	90.7
		Olteanu F.		5.28	26.44	35.64	47.96	57.64	21	1.76	92.1
		Petrariu C.		5.26	26.48	35.79	48.39	58.35	19	1.15	91.0
							Total	3:51.89			
21	19	Peterson L.	NZL-I	5.36	26.68	35.93	48.26	57.98	23	1.82	91.5
		Telford B.		5.36	26.83	36.16	48.62	58.42	18	1.14	90.8
		Dacre R.		5.39	26.48	35.59	47.73	57.30	17	1.42	92.6
		Henry P.		5.40	26.59	35.96	48.66	58.67	21	1.47	90.5
							Total	3:52.37			
22	1	Chen C.	TPE-I	5.33	26.55	35.66	47.69	57.03	10	0.87	93.1
		Chen C.		5.34	26.96	36.36	49.02	58.94	23	1.66	90.0
		Lee C.		5.35	26.52	35.76	48.23	57.98	23	2.10	91.5
		Wang J.		5.33	26.57	35.91	48.73	58.80	23	1.60	90.3
							Total	3:52.75			
23	25	Di iazza A.	AUS-I	5.54	27.03	36.29	48.58	58.20	25	2.04	91.2
		Harland M.		5.56	27.21	36.57	49.07	58.87	22	1.59	90.1
		Dodd S.		5.57	26.84	35.91	47.87	57.75	16	1.37	92.7
		Craig S.		5.53	26.94	36.32	49.03	59.02	24	1.82	89.9
							Total	3:53.34			
24	8	Victorov T.	BUL-I	5.51	26.85	36.01	48.14	57.72	21	1.56	91.9
		Stamov P.		5.52	27.28	36.65	49.16	59.07	24	1.79	89.8
		Botev N.		5.49	26.92	36.14	48.48	58.20	24	2.32	91.2
		Simeonov A.		5.53	26.89	36.17	48.74	58.67	21	1.47	90.5
							Total	3:53.66			
25	5	Reis A.	POR-I	5.65	27.34	36.58	48.83	58.42	26	2.26	90.8
		Poupada J.		5.67	27.73	37.15	49.77	59.67	26	2.39	88.9
		Pires J.		5.64	27.08	36.32	48.62	58.28	25	2.40	91.1
		Bernardes R.		5.67	27.23	36.56	49.16	59.13	25	1.93	89.8
							Total	3:55.50			
	18	Stokes D.	JAM-I	5.34	26.95	36.19	48.44	58.04	24	1.88	91.4
		Harris D.		5.35	27.43	36.87	49.46	59.37	25	2.09	89.4
		White M.									
		Stokes N.									

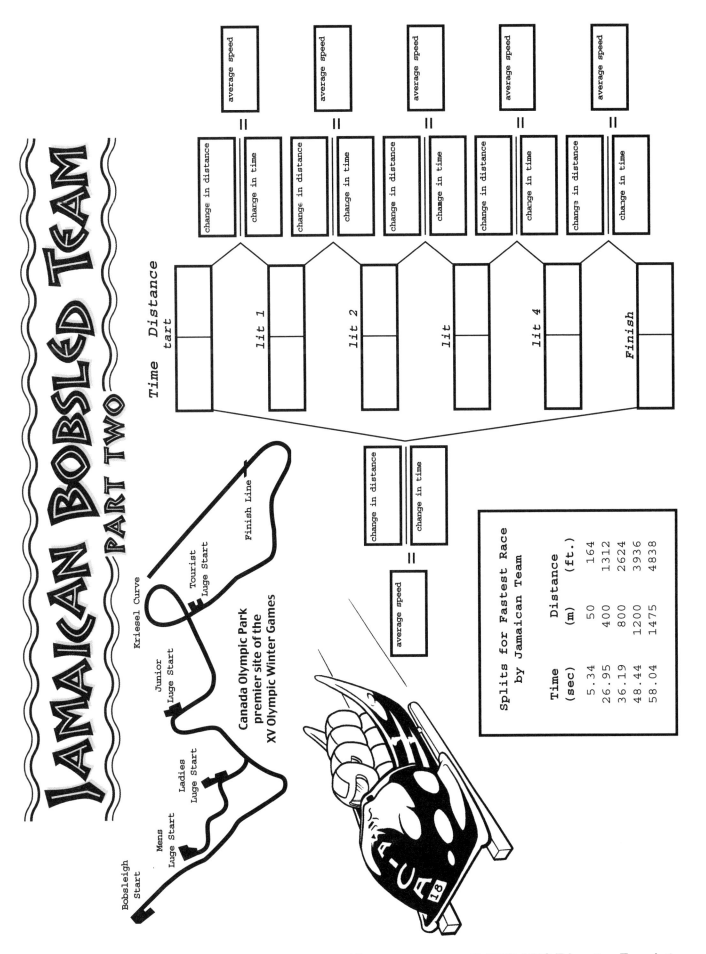

JAMAICAN BOBSLED TEAM
PART TWO

Kriesel Curve

Finish Line

Tourist
Luge Start

Junior
Luge Start

Ladies
Luge Start

Mens
Luge Start

Bobsleigh
Start

**Canada Olympic Park
premier site of the
XV Olympic Winter Games**

Time Distance
Start

Lap 1

Lap 2

Lap

Lap 4

Finish

change in distance
change in time
=
average speed

change in distance
change in time
=
average speed

change in distance
change in time
=
average speed

change in distance
change in time
=
average speed

change in distance
change in time
=
average speed

change in distance
change in time
=
average speed

Splits for Fastest Race by Jamaican Team		
Time (sec)	Distance (m)	(ft.)
5.34	50	164
26.95	400	1312
36.19	800	2624
48.44	1200	3936
58.04	1475	4838

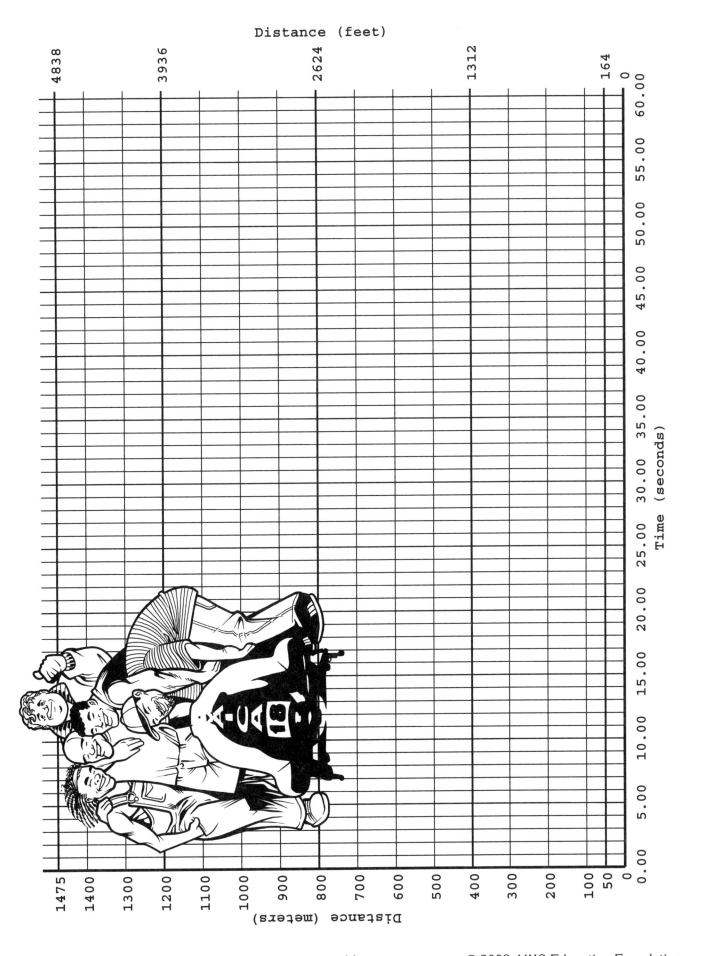

Distance (feet)

4838   3936   2624   1312   164   0

0.00   5.00   10.00   15.00   20.00   25.00   30.00   35.00   40.00   45.00   50.00   55.00   60.00

Time (seconds)

1475   1400   1300   1200   1100   1000   900   800   700   600   500   400   300   200   100   50   0

Distance (meters)

THE HUNT FOR RED OCTOBER

Captain Marko Ramius is the commander of a new Typhoon-classed Soviet submarine. This high-tech nuclear sub has an advanced new caterpillar drive that creates plenty of nervousness among the brass of the U.S. military. Captain Ramius causes the entire Russian naval fleet to engage in a large-scale hunt as he navigates the Red October toward the United States. Is Captain Ramius to use the Red October to defect to the U.S.? If defecting is on his mind, how will he convince the entire crew to support him in this deadly endeavor?

In the movie *The Hunt for Red October*, Captain Ramius evades his pursuers by making a high speed run through an underwater canyon. During this segment of the movie, the viewer hears the course heading, speed, and duration of each leg of the run through the safe lane. Everything is very precisely measured because an error would send the submarine into a collision with the canyon walls.

This movie shows sailors using their math in action. Using their understanding of geometry and rates, they plot a safe course through the canyon.

With the information in the movie, the students can reconstruct the map and develop instructions of course and headings for a safe lane.

The PG rating of this movie makes it acceptable to most students. There are some violent scenes and inappropriate language, but none appears in the segments suggested for student viewing.

RUNNING THE LANE

Topic
Rates and scale maps

Key Question
How can you take the course heading, speed, and duration of each leg of the submarine running the lane to determine the shape of the canyon and the safe lane that runs through it?

Learning Goals
Students will:
1. learn to determine distance given rate and time, and
2. learn to construct scale maps given distance and headings.

Guiding Documents
Project 2061 Benchmarks
- *Mathematics is helpful in almost every kind of human endeavor—from laying bricks to prescribing medicine or drawing a face. In particular, mathematics has contributed to progress in science and technology for thousands of years and still continues to do so.*
- *Estimate distances and travel times from maps and the actual size of objects from scale drawings.*
- *Use calculators to compare amounts proportionally.*

NCTM Standards 2000*
- *Develop, analyze, and explain methods for solving problems involving proportions, such as scaling and finding equivalent ratios*
- *Solve problems involving scale factors, using ratio and proportion*
- *Select and apply techniques and tools to accurately find length, area, volume, and angle measures to appropriate levels of precision*
- *Recognize and apply geometric ideas and relationships in areas outside the mathematics classroom, such as art, science, and everyday life*

Math
Proportional reasoning
    rates
    scaling
Measurement
    angle

Science
Earth science
    orienteering

Integrated Processes
Observing
Collecting and organizing data
Predicting
Inferring
Interpreting data

Materials
Overhead transparency film
Scissors
Straight edge

Background Information
In the movie *The Hunt for Red October*, the Russian submarine evades its pursuers by making a high speed run through an underwater canyon. During this segment of the movie, the viewer hears the course heading, speed, and duration of each leg of the run through the safe lane. Everything is very precisely measured because an error would send the submarine into a collision with the canyon walls. Enough data is provided to reconstruct a map of the safe lane through the canyon.

Speed is given in knots, which means nautical miles per hour. To determine the distance traveled, students will need to use the equation $d = rt$, where d represents distance traveled, r represents the rate or speed of travel, and t represents the duration of time traveled. To determine the distance traveled during each leg, the duration of the leg is multiplied by the speed. The speed is nautical miles per hour so the time needs to be in hours. In the movie, the time is given in minutes so it must be converted to fractions of an hour, or a fraction of minutes over 60 minutes per hour. The calculations for the first three legs are given below.

	Duration	Hours	Speed	Distance
Leg 1:	6.5 min.	6.5/60 hrs.	18 knots	1.95 nautical miles
Leg 2:	8.67 min.	8.67/60 hrs.	26 knots	3.76 nautical miles
Leg 3:	34 min.	34/60 hrs.	26 knots	14.73 nautical miles

To make a scaled map of the lane, a protractor is used to set the direction of the course relative to north. The submarine would travel on this heading the calculated distance, which can be represented by a scaled line. Using a scale of one cm to one nautical mile makes it quite easy to convert the measurements into their scaled distances.

At the end of the third leg, the crew becomes very concerned they will collide with "Neptune's Massive," a large rock formation in front of them. The captain has ordered them to maintain their course for 15 to 30 seconds past the critical turn to evade a torpedo.

To determine how far they are out of the safe lane, one can calculate the distance. However, for this distance to be comprehensible, nautical miles must be converted to a more familiar measure. A nautical mile is 6076.1 feet, or 1852 meters. Knowing this, the distance in nautical miles can be multiplied by the appropriate conversion factor.

Duration	Hours	Speed	Distance
15 sec.	0.25/60 hrs.	26 knots	0.1083 n.m. = 658.24 ft. = 200.63 m
30 sec.	0.5/60 hrs.	26 knots	0.216 n.m. = 1316.27 ft.= 401.27 m

The last two legs of the run are not shown in the movie. But given a map of the lane, the duration of the legs can be determined in the same way they would have been by the navigator on the submarine. The two legs measure 2.8 cm (2.8 n.m.) and 2.1 cm (2.1 n.m.) Setting up a ratio of distance to speed, the duration can be determined in hours. Multiplying the duration in hours by 60 determines the duration in minutes.

Distance	Speed	Duration
2.8 n.m.	26 knots	0.1077 hours = 6.46 min. = 6:28 min:sec
2.1 n.m.	26 knots	0.0808 hours = 4.85 min. = 4:51 min:sec

Management

1. Every student will need one of the 360° protractors provided for this activity. These protractors will need to be copied onto overhead transparency film before use. Some instruction will be needed for students to use the protractors. Make sure the protractors are facing right side up. Have the students notice that these protractors have the 360° of a compass face. To get accurate measurements, the students will need to make sure the protractors are aligned with the map. The center of the protractor should be directly on top of the position being measured. The lines of the protractor's center grid should align with the grid of the map and north, or 0°, should be towards the top of the map.
2. A map of the canyon and safe lane is provided so students can check their accuracy. When students complete their drawing of the submarine's run, they can place their drawing on top of the map. By holding both sheets towards a light and aligning the grids of the drawing and the map, they can see how well they stayed in the safe lane.
3. To provide the most involvement by students in this activity, the segment from *The Hunt for Red October* should be shown to the class. Before class, find the segment near the middle of the movie where the submarine is running the lane from "Thor's Twins" to evading the torpedo near "Neptune's Massive." Prepare the segment so it is ready to show the class.

Procedure

1. Show the segment of *The Hunt for Red October* and have the students discuss the *Key Question*.
2. Distribute the first student sheet. Have students record the duration of each leg as a fraction of an hour and then calculate the distance traveled.
3. Distribute the blank grid page and protractors. Make sure the students know how to measure with the protractor.
4. Using the data from the chart, have students make a scaled drawing of the submarine's run through the lane.
5. When the students have completed their drawings for the first three legs, have them check their drawings with the map.
6. Have students use the map to make measurements for the last two legs and enter the data on their charts. They will need to work backwards to determine the duration.
7. Have students discuss the accuracy of their drawings and how they could improve them in the future.

Discussion

1. How can you convert minutes to the equivalent hours? [ratio, ex.: 15 min/60 min per hour = 1/4 hour = 0.25 hours]
2. If you know the time (t) you been traveling and the rate (r) or speed at which you have been traveling, how do you determine the distance (d) you have traveled? [d = r • t]
3. If you know the distance you need to travel and the rate or speed you are traveling, how do you determine the time it will take you to travel? [work backwards if d = r • t then d ÷ r = t]
4. How could you improve you drawings to be more accurate?
5. When might you use a skill like this? [navigation, orienteering, looking for buried treasure, giving directions]

Extension

Have the students use the information from the movie to determine how far out of the lane the submarine got as it stayed on course (see *Background Information*).

* Reprinted with permission from *Principles and Standards for School Mathematics*, 2000 by the National Council of Teachers of Mathematics. All rights reserved.

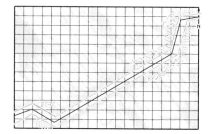

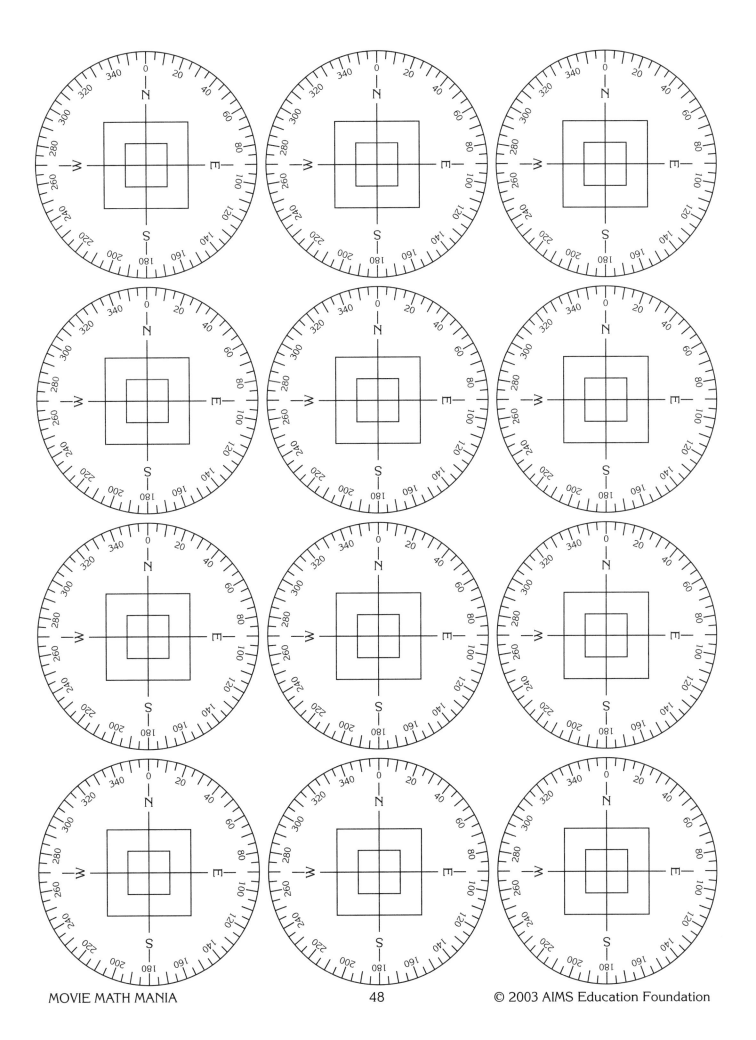

RUNNING THE LANE

Directions for lane run from Thor's Twins

Landmark	Duration		Course	Speed	Distance
	min:sec	min/60 hours			

Thors Twins: Commence Run

Landmark	min:sec	min/60 hours	Course	Speed	Distance
Leg 1	6:30		260°	18 knots	
Leg 2	8:40		195°	26 knots	
Leg 3	34:00		240°	26 knots	

Neptune's Massive

Landmark	min:sec	min/60 hours	Course	Speed	Distance
Leg 4				26 knots	
Leg 5				26 knots	

RUNNING THE LANE

1 cm: 1 nautical mile

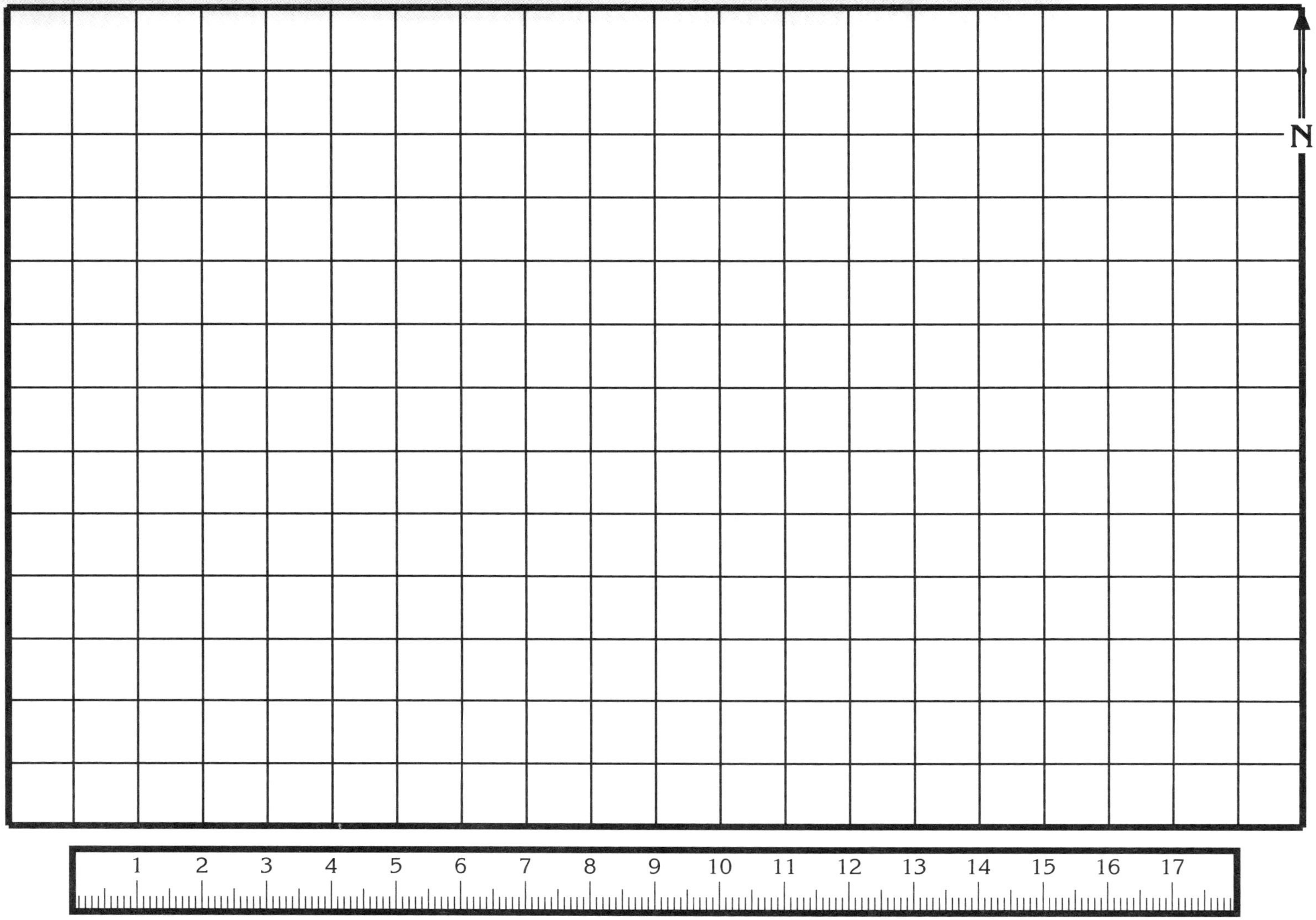

N

RUNNING THE LANE

1 cm: 1 nautical mile

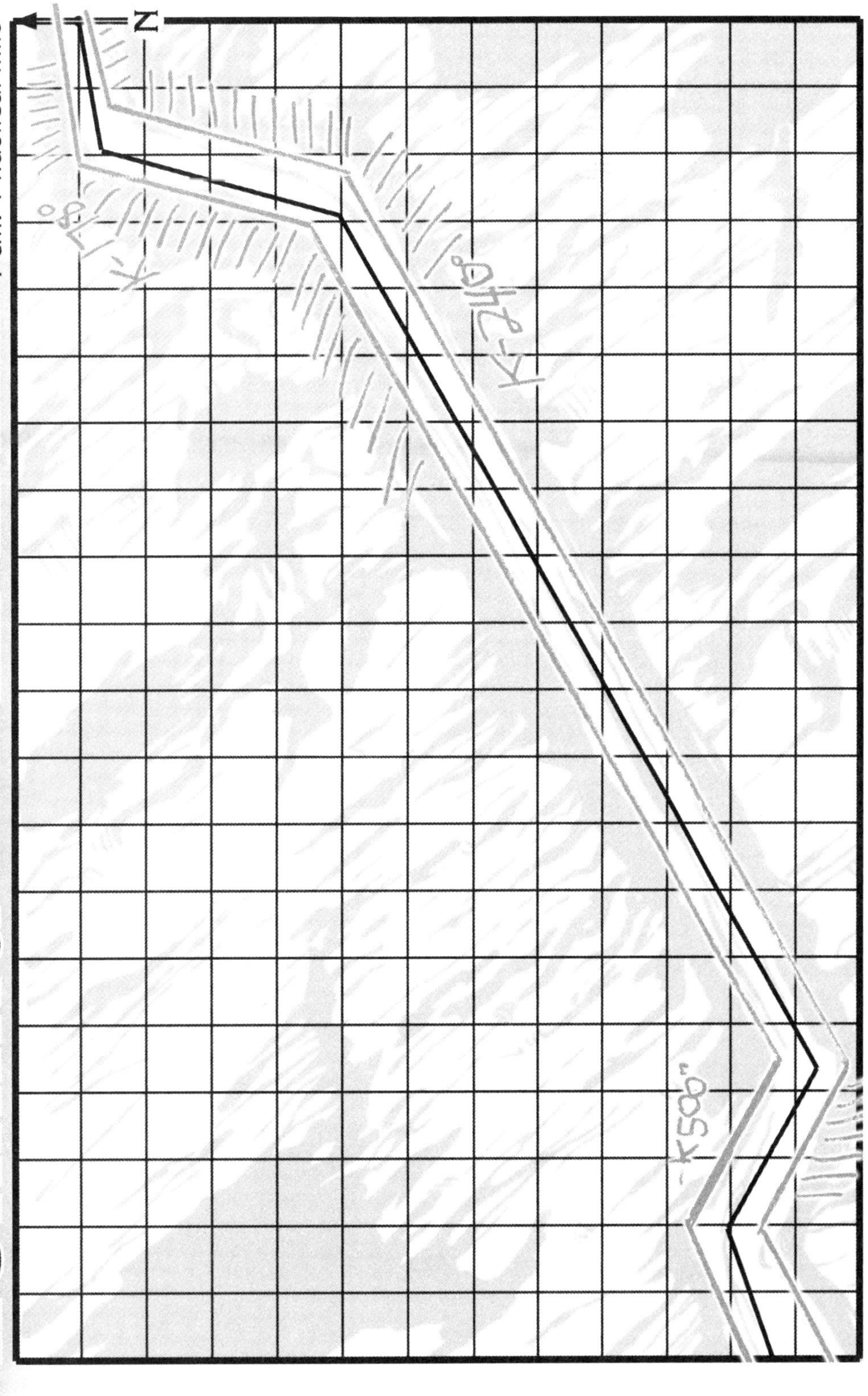

N

K-130°

K-210°

-K 500"

TAKING COMMAND

Topic
Rates and scale maps

Key Question
How can you give directions to guide a submarine on a safe course through enemy infested waters?

Learning Goals
Students will:
1. determine time given distance and rate, and
2. ather course and distance data from scaled maps.

Guiding Documents
Project 2061 Benchmarks
- *Mathematics is helpful in almost every kind of human endeavor—from laying bricks to prescribing medicine or drawing a face. In particular, mathematics has contributed to progress in science and technology for thousands of years and still continues to do so.*
- *Estimate distances and travel times from maps and the actual size of objects from scale drawings.*
- *Use calculators to compare amounts proportionally.*

NCTM Standards 2000*
- *Develop, analyze, and explain methods for solving problems involving proportions, such as scaling and finding equivalent ratios*
- *Solve problems involving scale factors, using ratio and proportion*
- *Select and apply techniques and tools to accurately find length, area, volume, and angle measures to appropriate levels of precision*
- *Recognize and apply geometric ideas and relationships in areas outside the mathematics classroom, such as art, science, and everyday life*

Math
Proportional reasoning
    rates
    scaling
Measurement
    angle

Science
Earth science
    orienteering

Integrated Processes
Observing
Comparing and contrasting
Collecting and organizing data
Predicting
Inferring
Interpreting data

Materials
Transparency protractor (see *Management 1*)
Scissors
Centimeter rulers

Background Information
This activity is an application and modification of what is learned in the activity *Running the Lane* based on the movie *Hunt for Red October*.

In this activity, each student draws a safe route for a submarine through an area of the sea where the enemy is present. Using the protractor and scaling distances with a ruler, students gather data about their route. They choose the speed the submarine should go and then work backwards to determine the duration of each leg of their route in minutes and seconds.

The accuracy of the directions can be checked by using a blank grid and only the directions of duration, course, and speed. After making conversions to distance and using the scale and protractor a new drawing of the route is drawn. This can be laid over the original to check that it is the same.

Management
1. Every student will need a protractor for this activity. Before starting the activity, protractors will need to be copied onto overhead transparency film. Some instruction will be needed for students to use the protractors. Make sure the protractors are facing right side up. Have the students notice that these are the 360° protractors of a compass face. To get accurate measurements, the students will need to make sure the protractors are aligned with the map. The center of the protractor should be directly on top of the position being measured. The lines of the protractor's center grid should align with the grid of the map and the north or 0° should be towards the top of the map.
2. Before beginning the activity, decide if students will draw their own naval situations, or will use the prepared battle map. Students should be provided with a blank grid if they are to design their own battle map.

3. This investigation assumes students have done the prior activity *Running the Lane.*
4. This activity works well over a two-day span. On the first day, students will draw their routes, gather data, and convert the data into directions. The second day the directions for duration, course, and speed will be followed to check their accuracy.

Procedure
1. Show a sample of the battle map to the students and ask them to consider the *Key Question.*
2. If students are to design their own battle maps, distribute a blank grid to each student. Let them place enemy ships and submarines as well as geographic objects on the map. Warn them not to make the objects so close that there is no path through.
3. On their battle map, direct each student to draw a route through the enemy and objects to the opposite side using only straight lines.
4. Have each student measure and record the course and distance data for each leg of the route. For each route, ask them to make up a speed between 0 and 30 knots. Inform them that slower speeds create less noise.
5. Using the data, have each student calculate the duration of each leg in minutes and seconds.
6. Tell them to copy the duration in minutes and seconds, the course, and the speed of their route onto a blank chart.
7. Have students exchange their partial set of directions and calculate the distance of each leg.
8. On a blank grid, instruct students to draw the route following the course and distance data.
9. Have them compare their drawn routes with the original maps for accuracy by placing one sheet on top of the other and looking towards a light.

Discussion
1. If you know the distance (d) and rate you have been traveling, how do you determine the time (t) you have traveled? [d ÷ r = t]
2. How can you convert hours to equivalent minutes and seconds? [hours • 60 = minutes, take the decimal value of the minutes times 60 to determine the seconds]
3. How could you improve your drawings to be more accurate?
4. When might you use a skill like this? [navigation, orienteering, looking for buried treasure, giving directions]

*   Reprinted with permission from *Principles and Standards for School Mathematics,* 2000 by the National Council of Teachers of Mathematics. All rights reserved.

TAKING COMMAND

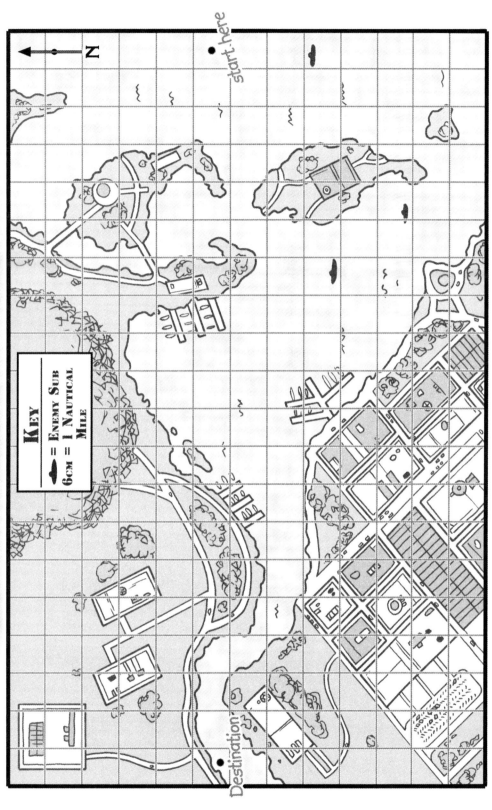

start here

N

KEY
= ENEMY SUB
6CM = 1 NAUTICAL MILE

Destination

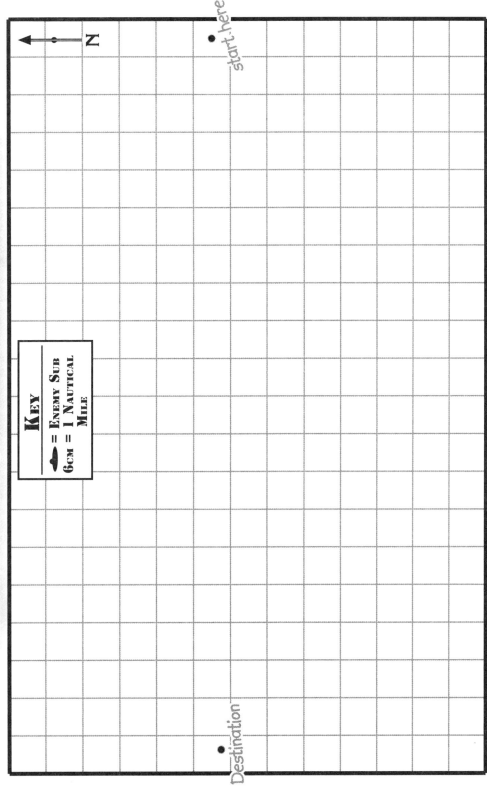

TAKING COMMAND

Key

➤ = Enemy Sub

6CM = 1 Nautical Mile

N

start here

Destination

TAKING COMMAND

LISTEN TO THE GIRL.

Directions for a safe run:

Leg of Trip	Duration		Course degrees	Speed knots	Distance nautical miles
	min:sec	hours			
Leg 1					
Leg 2					
Leg 3					
Leg 4					
Leg 5					
Leg 6					
Leg 7					
Leg 8					

The movie U-571 is a submarine action film about World War II. Hitler's U-boats are wreaking havoc in the Atlantic and destroying Allied shipping up and down the United States' East Coast. Unable to crack the U-boat radio codes, the U.S. struggles blindly against the Germans, until an American submarine crew is ordered to capture the enigma machine, the German's top-secret coding device. The American submarine, the S-33, is rigged to resemble a Nazi U-boat and is sent out as a rescue ship. If all goes according to plan, the crew will take the Germans as prisoners and steal the enigma machine. The plan fails, and nine U.S. soldiers are trapped in a submarine in the middle of hostile enemy waters fighting to survive.

This exciting context provides an opportunity to look at rates and the algebraic representations as graphs and equations of linear functions. By studying the gauges in the movie, students determine depth and ascent rates. Using what data they gather from the movie, the students use algebraic thinking to determine how much time the crew has to prepare to fight the enemy on the surface. As the movie continues, students can check the accuracy of their predictions.

U-571 is rated PG-13 for war violence and language. The last 20 minutes, which are related to the activities have very little of either.

U-571 | PART ONE
HOW DEEP DID THE SUBMARINE GO?

Topic
Algebra—Linear Functions

Key Questions
How can you determine the greatest depth the submarine dove by knowing how much the depth gauge rotated?

Learning Goals
Students will:
1. measure the angle of rotation on the depth gauge at different depths, and
2. graph the data to identify the proportional or linear nature of the data.

Guiding Documents
Project 2061 Benchmarks
- *The graphic display of numbers may help to show patterns such as trends, varying rates of change, gaps, or clusters. Such patterns sometimes can be used to make predictions about the phenomena being graphed.*
- *Mathematical statements can be used to describe how one quantity changes when another changes. Rates of change can be computed from magnitudes and vice versa.*

NRC Standard
- *Use appropriate tools and techniques to gather, analyze, and interpret data.*

NCTM Standards 2000*
- *Understand and use ratios and proportions to represent quantitative relationships*
- *Relate and compare different forms of representation for a relationship*
- *Use symbolic algebra to represent situations and to solve problems, especially those that involve linear relationships*
- *Model and solve contextualized problems using various representations, such as graphs, tables, and equations*

Math
Measurement
   angular
Proportional reasoning
Algebra
   linear functions
   graphing
   equations

Science
Physical science
   pressure

Integrated Processes
Observing
Collecting and organizing data
Interpreting data
Generalizing

Materials
Overhead transparency copier film (or protractors)

Background Information
This is the first in a series of three investigations based on the movie *U-571*. This realistic movie of a World War II submarine battle provides a highly motivating context in which to study linear functions.

In the movie, a submarine dives to a depth of 200 meters and then begins to sink out of control. Shots of the depth gauge are the only accurate measure of depth in the movie. With no scale on the gauge below 200 meters, the only way to estimate the final depth is to determine the relation between the angle of rotation and the depth. Students measure the three shots of the gauge shown in the movie and graph the data. A straight line that passes through the origin alerts the students that the data are proportional and may be written as a linear function. Dividing the degrees of rotation by the depth tells how many degrees of rotation there are for each meter of depth ($1.35°/m$). This ratio is the slope of the graph. ($1.35°$ up for each 1 meter to the right.) Written in function notation where the angle depends on the depth ($A(d)$), this ratio becomes the coefficient with the variable for depth(d). ($A(d)=1.35d$)

By extending the line on the graph to 318° of rotation allows one to estimate the final depth to be in the range of 230-240 meters. A more accurate prediction can be made by substituting the final angle of rotation for A(d). Then by working the equation backwards, one can solve for the depth (d):

$$A(d) = 1.35d$$
$$318 = 1.35d$$
$$318/1.35 = d$$
$$235.5 = d$$

Management

1. This activity is much more motivating for students if they have seen the movie. Encourage them to view the movie at home or have a "party" and show the movie outside of school time. Borrow, rent, or purchase the VCR or DVD version of *U-571* and find the equipment to show the appropriate clips at school.
2. Before beginning this activity, copy the protractors on overhead transparencies so there is at least one for each group of students.

Procedure

1. Distribute the protractors and the page with the four pictures of the depth gauge. Then pose the *Key Question* to the students.
2. Have the students measure and record the angle of rotation for each of the gauge pictures. Some supervision or instruction may be needed if students are not familiar with the use of a protractor. Make sure the protractors are right side up, the 0° marker is aligned with the zero on the scale of the gauge, and the center of the protractor is on the center of the gauge (the circle around the protractor should overlay the outside of the gauge).
3. Using their measurements, have the students plot the data on the graph.
4. Have students discuss what patterns the data form on the graph [forms a line] and how they could use the graph to predict the final depth [extrapolate line to known angle of rotation].
5. Ask the students to consider their numeric data and determine a way to calculate the final depth from the known angle.
6. Have students develop an equation that will calculate the depth when the angle of rotation is substituted into the equation.

Discussion

1. What happens to the gauge as the submarine descends? [rotates, turns]
2. The amount the gauge turns is caused by what change? [how far the submarine goes down]
3. What pattern do you see in the points of data on your graph? [form a line that goes up as you go to the right]

4. How could you use your graph to determine the depth of the submarine when the picture of the last gauge was made? [extend line until it reached the last angle of rotation]
5. How could you get a more precise answer by using the measurements you made? [proportion, unit rate with equation, depth 235 meters]
    5a. From your data, can you determine how many degrees the pointer turned for the amount of meters it went down? [12°/9m = 28°/21m = 220°/163m 1.35°/1m = 1.35°/m]
    5b. How could you use this ratio or relationship to determine the final depth of the submarine? [220°/163m = 318°/x or 318° ÷ 1.35°/m = x m]
6. How would you write an equation that calculates the depth (d) of the submarine from the angle of rotation(r)? [r = 1.35d , d = r/1.35]

*   Reprinted with permission from P*rinciples and Standards for School Mathematics, 20*00 by the National Council of Teachers of Mathematics. All rights reserved.

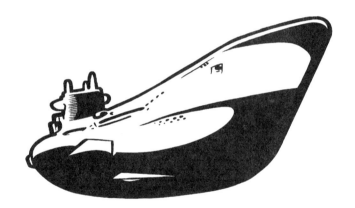

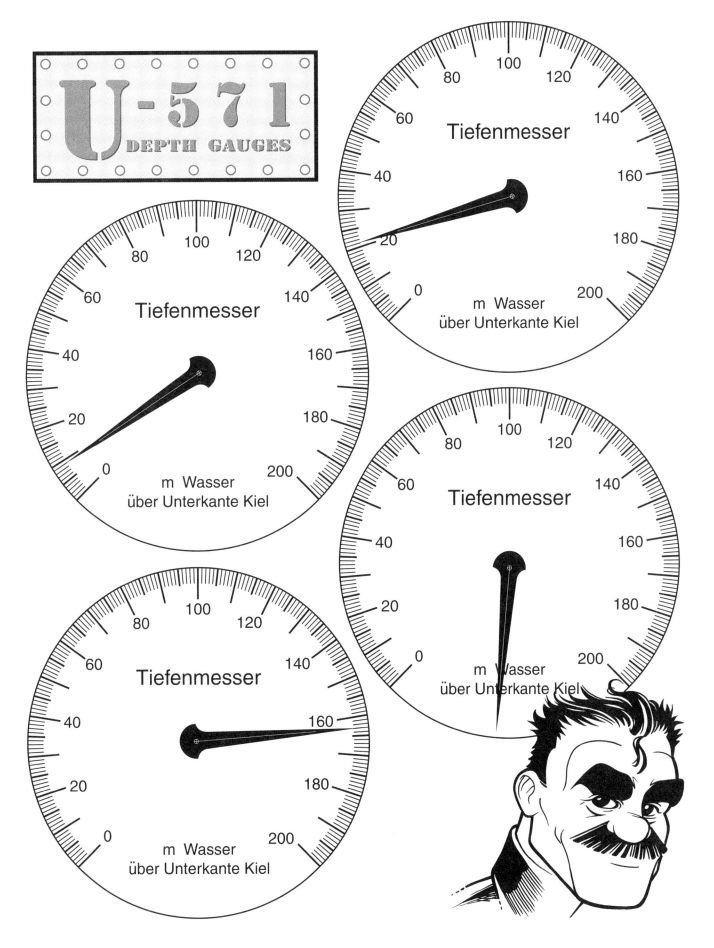

61

U-571

HOW DEEP DID THE SUBMARINE GO?

CAN YOU DETERMINE THE GREATEST DEPTH THE SUBMARINE DOVE BY KNOWING HOW MUCH THE DEPTH GAUGE ROTATED?

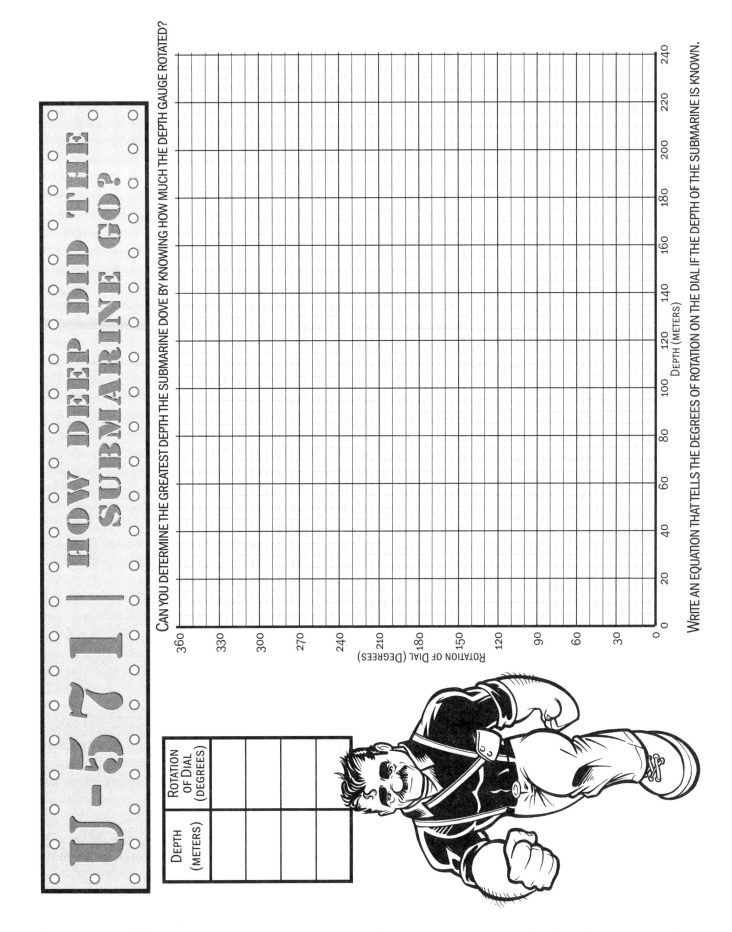

DEPTH (METERS)	
ROTATION OF DIAL (DEGREES)	

ROTATION OF DIAL (DEGREES)

DEPTH (METERS)

WRITE AN EQUATION THAT TELLS THE DEGREES OF ROTATION ON THE DIAL IF THE DEPTH OF THE SUBMARINE IS KNOWN.

Topic
Algebra—Linear Functions

Key Questions
How much pressure was there at the final depth of the submarine?

How can you determine the depth of a submarine if you know the pressure where it is?

Learning Goals
Students will:
1. measure the pressure at different depths, and
2. graph the data to identify the linear relationship of pressure to depth.

Guiding Documents

Project 2061 Benchmarks
- *The graphic display of numbers may help to show patterns such as trends, varying rates of change, gaps, or clusters. Such patterns sometimes can be used to make predictions about the phenomena being graphed.*
- *Mathematical statements can be used to describe how one quantity changes when another changes. Rates of change can be computed from magnitudes and vice versa.*

NRC Standard
- *Use appropriate tools and techniques to gather, analyze, and interpret data.*

NCTM Standards 2000*
- *Understand and use ratios and proportions to represent quantitative relationships*
- *Relate and compare different forms of representation for a relationship*
- *Use symbolic algebra to represent situations and to solve problems, especially those that involve linear relationships*
- *Model and solve contextualized problems using various representations, such as graphs, tables, and equations*

Math
Algebra
    linear functions
    graphing
    equations

Science
Physical science
    pressure

Integrated Processes
Observing
Collecting and organizing data
Interpreting data
Generalizing

Materials
Student pages

Background Information
This is the second in a series of three investigations based on the movie *U-571*. This realistic movie of a World War II submarine battle provides a highly motivating context in which to study the linear functions.

The pressure on a submarine increases at a constant rate as it descends into the ocean. The combined force exerted by the column of water above the submarine and the pressure of the atmosphere causes pressure. The pressure of the atmosphere fluctuates because of various meteorological causes, but for this activity 1.03 km/cm² has been used. Likewise the pressure of the water changes due to temperature and salinity, but an average of an increase of 0.103 kg/cm² per meter of depth is typical. The pressure on a submarine is a function of its depth. One could say that the pressure on a submarine depends on its depth.

One way to get an idea of a function's nature is to graph some of the dependent variable (F(x) or y) data against the independent variable (x). When the three data points of pressure and depth are graphed, they can be connected by a straight line thereby demonstrating the linear nature of the function. The rate at which the pressure increases can be determined by dividing the total change in pressure by the total change in depth.

$$\frac{\Delta y}{\Delta x} = \frac{y_2 - y_1}{x_2 - x_1} = \frac{17.8 - 2.0}{163 - 9} = \frac{15.8}{154} \approx 0.103 \ \frac{kg/cm^2}{m}$$

The pressure increased about 0.103 kilograms per square centimeter every meter the submarine descends.

Looking at the graph, the line does not show zero pressure when the depth is zero meters. There must be some pressure before the submarine descends. Looking at the y-intercept on the graph gives an approximation of 1 kg/cm² of pressure. This is can be calculated in several ways. Most students begin repeatedly subtracting 0.103 kg/cm² from the pressure at nine meters of depth (2.0 kg/cm²). Some students will suggest multiplying 0.103 kg/cm² by nine to find the change and then subtract the total change from 2.0 kg/cm². For those students familiar with the slope-intercept form (y=mx+b), the teacher might suggest substituting the y, m, and x with known quantities, and then solve for the unknown.

$$y=mx+b$$
$$2.0=(0.103)(9)+b$$
$$2.0-(0.103)(9)=b$$
$$b = 1.073 \doteq 1.1$$

The approximate value of 1.1 kg/cm² is close to atmospheric pressure. The difference is caused by the precision of the numbers from the gauges.

Since pressure (P(d)) is a function of depth in meters(d), an equation can be developed from the understanding that students have gotten by doing the calculations. Multiplying the number of meters by the increase in pressure tells the amount of pressure from the water. By adding the atmospheric pressure, you determine the total pressure on the submarine.

$$P(d) = 0.103d + 1.1$$

As students are asked to compare the numeric, graphic, and symbolic representations of the data, they will generalize that the change in pressure shows up as the slope of the line and the coefficient in the equation. The atmospheric pressure is then added to the equation and is the y intercept on the graph.

Management

1. This activity is much more motivating for students if they have seen the movie. Encourage them to view the movie at home or have a "party" and show the movie outside of school time. Borrow, rent, or purchase the VCR or DVD version of *U-571* and find the equipment to show the appropriate clips at school.
2. This investigation is for students who are becoming familiar with linear function ideas and can deal with the more difficult numbers. It provides an opportunity to see an application of multiple representations of linear function ideas and to begin to consider how to develop equations from partial data.

Procedure

1. Distribute the page with the four pictures of the depth and pressure gauges. Then pose the *Key Question* to the students.

2. Have the students measure and record the pressure and depth of each set of gauge pictures. Emphasize to students that they need to determine the increments on the scales of both gauges. [depth—1 m, pressure—0.2 kg/cm²]
3. Using their measurements, have the students plot the data on the graph.
4. Have students discuss what patterns the data form on the graph [form a line] and how they could use the graph to predict the final depth [extrapolate line].
5. Ask the students to consider their numeric data and determine a way to calculate the rate of pressure increase.
6. Have students discuss how to use the graph and the data to determine the pressure when the submarine is on the surface.
7. Using the rate of increase and initial pressure on the surface, have the students discuss how to determine the total pressure on the submarine as the depth changes and how to translate their instructions into an equation.
8. Have students substitute the pressure off the last gauge and work backwards to determine the submarine's final depth.

Discussion

1. What pattern do you see in the points of data on your graph? [form a line that goes up as you go to the right]
2. How can you determine how much the pressure increases each time the submarine descends a meter? (See B*ackground Information*.)
3. According to your graph, what is the pressure on the submarine when it is on the surface (at a depth of zero meters)? [about 1 kg/cm²]
4. How could you use your data to get a more precise measurement of the pressure at the surface? (See *Background Information*.)
5. How can you determine the total pressure on the submarine? [multiply the depth(m) by the increase per meter (0.103) and add the initial pressure at the surface (1.1)]
6. What would your instructions for finding the total pressure (P(d)) on the submarine look like? [P(d) = 0.103d + 1.1]
7. Where do the coefficient and constant in the equation show up on the graph? [slope, y-intercept]
8. How could you determine the final depth of the submarine since you only know the pressure? [substitute 25.3 in for P(d) and solve for d. (See *Background Information*.)]

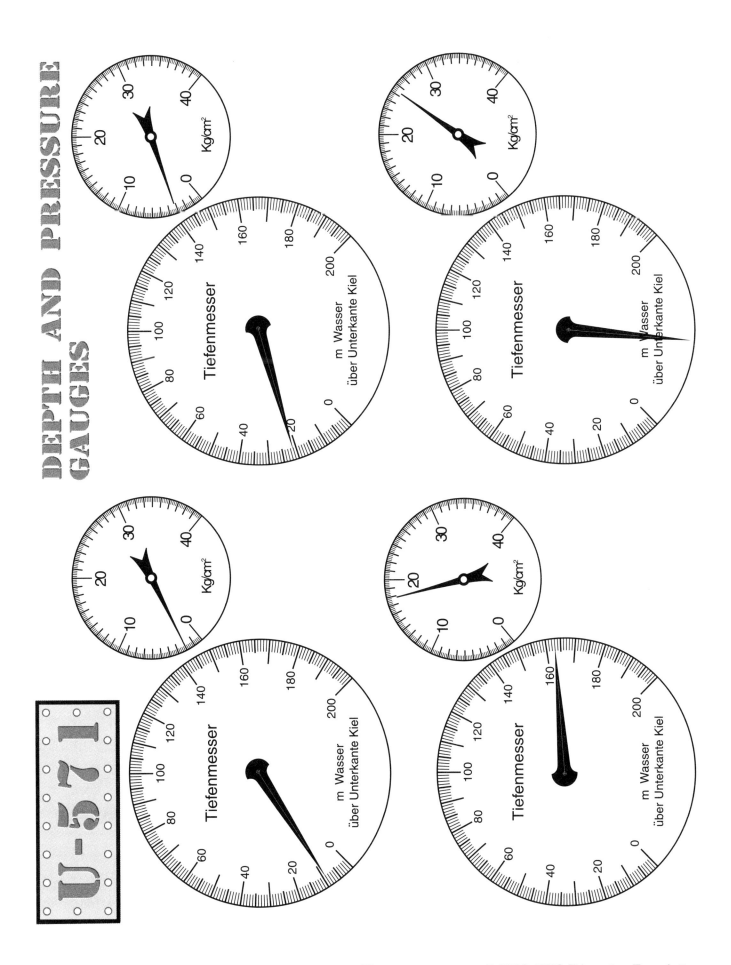

U-5-7 1 2-5-1

HOW DEEP DID THE SUBMARINE GO?

CAN YOU DETERMINE THE GREATEST DEPTH THE SUBMARINE DOVE BY KNOWING HOW MUCH THE DEPTH GAUGE ROTATED?

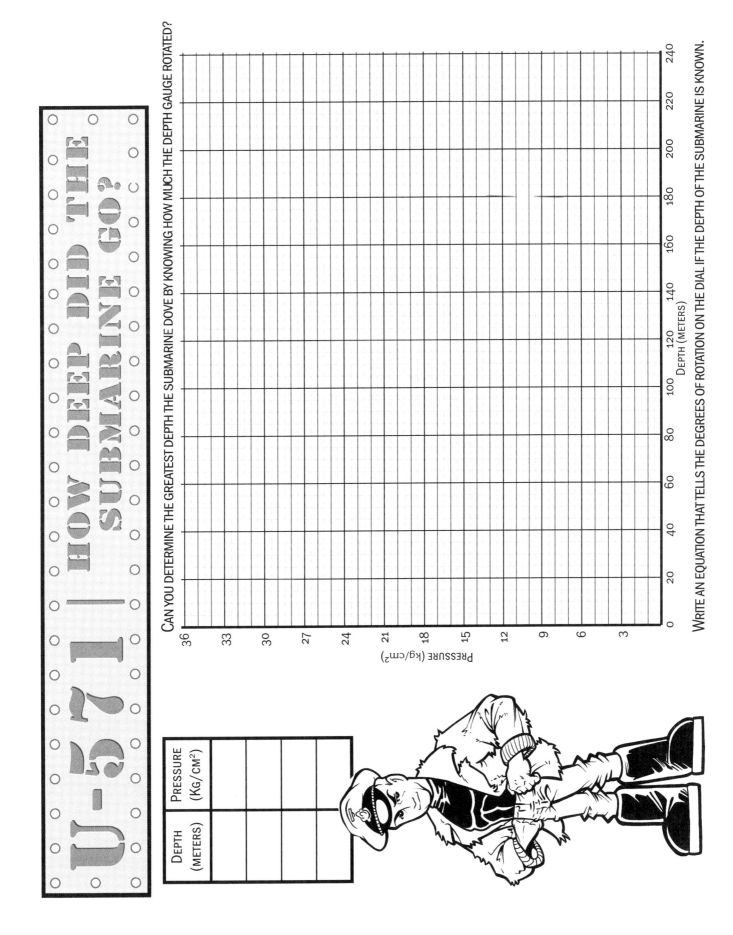

DEPTH (METERS)	PRESSURE (Kg/cm²)

WRITE AN EQUATION THAT TELLS THE DEGREES OF ROTATION ON THE DIAL IF THE DEPTH OF THE SUBMARINE IS KNOWN.

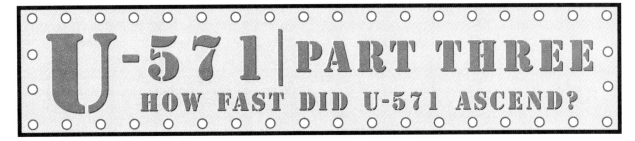

Topic
Algebra—linear functions

Key Question
Using the two sequences of the depth gauge, how could you estimate how much time the U-571 has until it would surface?

Learning Goals
Students will:
1. view two sequences of a depth gauge to determine the rate of ascent and how much time the total ascent would take, and
2. watch the whole ascent graphing the depth against time to determine how accurately the movie was made.

Guiding Documents
Project 2061 Benchmarks
- *The graphic display of numbers may help to show patterns such as trends, varying rates of change, gaps, or clusters. Such patterns sometimes can be used to make predictions about the phenomena being graphed.*
- *Mathematical statements can be used to describe how one quantity changes when another changes. Rates of change can be computed from magnitudes and vice versa.*

NRC Standard
- *Use appropriate tools and techniques to gather, analyze, and interpret data.*

NCTM Standards 2000*
- *Understand and use ratios and proportions to represent quantitative relationships*
- *Relate and compare different forms of representation for a relationship*
- *Use symbolic algebra to represent situations and to solve problems, especially those that involve linear relationships*
- *Model and solve contextualized problems using various representations, such as graphs, tables, and equations*

Math
Algebra
    linear functions
    graphing
    equations

Science
Physical science
    pressure

Integrated Processes
Observing
Collecting and organizing data
Interpreting data
Generalizing

Materials
Video or DVD of *U-571*
Stopwatch

Background Information
This is the final in a series of three investigations based on the movie *U-571*. This realistic movie of a World War II submarine battle provides a highly motivating context in which to study the linear functions.

The climax of the movie comes as the damaged U-571 ascends uncontrollably to the surface where an enemy ship waits. The crew must get the submarine ready to defend itself in a very short time. The skipper has an estimate of the length of time they have to prepare. The viewer can also determine an estimate by viewing two sequences showing the depth gauge for short periods of time as the boat ascends. The first sequence lasts 2.6 seconds as the submarine ascends through –154 to –150 meters. The boat rises 4 meters in 2.6 seconds, which is a rate of 1.5 meters per second. The second sequence has the submarine rising 7 meters (–97m to –90m) in 3.5 second or a rate of 2 meters per second. Students knowing from the previous activities that the final depth is –235m can start a graph at this point and draw a line with a slope the estimated rate of ascent. If students draw a line for both estimated rates, they will get two lines that will give them a range of time for when U-571 will surface.

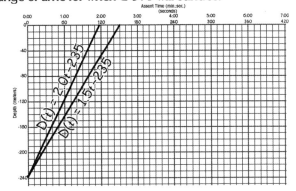

Students who have an understanding of linear functions might write the equations for both ascent rates and solve for the time for the submarine to reach a depth of zero.

$$D(t)=1.5t-235 \qquad \text{or} \qquad D(t)=2t-235$$
$$0=1.5t-235 \qquad\qquad\qquad 0=2t-235$$
$$235=1.5t \qquad\qquad\qquad\quad 235=2t$$
$$157t \qquad\qquad\qquad\qquad\quad 118t$$
$$2:37t \qquad\qquad\qquad\qquad\quad 1:58t$$

Using the depth gauge sequences, one would estimate that the crew had between two and three minutes to prepare. In fact the skipper confirms this estimate in the movie when they have just passed –90 meters and he tells Trigger they will be on the surface in one minute. However, the very long suspenseful segment seems much longer than three minutes. Did the producers make a mistake? By watching the whole ascent segment and graphing the depth to the elapsed time, a different line of ascent emerges. Below is a list of times in the segment when the depth is known.

Elapsed Time (m:s)	Depth(m)	Event
0:00	–235	Depth gauge shown
0:53	–200	Chief announces depth
2:34	–180	Chief announces depth
3:57	–150	Chief and gauge show depth
4:35	–90	Chief and gauge show depth
6:07	–30	Chief announces depth
6:20	–21	Depth gauge shown
6:28	–9	Depth gauge shown
6:35	0	Submarine shown surfacing

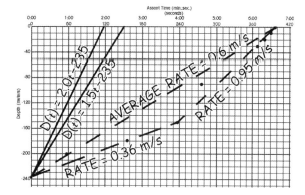

When looking at the points of the events on the graph, there appears to be two different slopes at different parts of the line. The slopes show two different ascent rates. In the first part of the ascent segment, the submarine rises 85 meters (–235 to –150) in 237 seconds (3:57) making the ascent rate 0.36 m/sec. At this rate, the crew would have ten minutes and fifty-three seconds until they surfaced.

The second part of the segment has the boat rise 150 meters (–150 m to 0 m) in 158 seconds or at a rate of 0.95 m/s. At this rate the submarine will be on the surface in four minutes and seven seconds. Although this is closer to the original estimate, it is significantly slower. Another approach is to determine the average rate over the whole ascent. The submarine rises 235 meters in 6 minutes and 30 seconds for a rate of 0.6 m/s (235m/390s = 0.6 m/s). This falls between the two ascent rates and is considerably slower than what was predicted.

The writer and director of *U-571* have taken creative license with the use of time and have taken advantage of the audiences' *willing suspension of disbelief*. In the plot of the story, the time seems consistent and reasonable. The glance at the depth gauge gives the viewer the sense that the submarine is ascending and increases the suspense. The viewer would not use it to analyze and estimate the time to surfacing. The director was not concerned at being true to time, but wanted to be true to the drama of the story.

Management
1. A VCR or DVD version of *U-571* is required for this investigation. Before doing the investigation, watch the segment of the movie concerned (the last 10–15 minutes of the video, or beginning in the middle of chapter 17 on DVD) and become familiar with the sequences showing the depth gauge.
2. This investigation is most appropriate for students who are familiar with linear function ideas and can deal with the more difficult numbers. It provides an opportunity to see an application of multiple representations of linear function ideas, how to develop equations from partial data, and dealing with the ambiguity of real data.

Procedure
1. Have the students watch one of the sequences where the depth gauge is shown moving in the movie *U-571* and have the class discuss how they could answer the first *Key Question*.
2. Have the students watch and time both sequences showing the moving depth gauge. Have them record the beginning and ending depths and time of the sequence on their record sheet. Have the students determine the ascent rate of each sequence.
3. Using the ascent rates they calculated and the starting depth of –235 m, have students make a graph of both ascent rates. They will have two different lines.
4. Have students discuss how they could use the graph to predict how much time it will take the submarine to ascend to the surface.
5. Using the graph and data, instruct the students to write equations for both ascent rates that calculate the depth when time is known.

6. Have the students use their equations to solve how much time is required to reach the surface for both rates.
7. Show the whole ascent segment. Start a stopwatch as soon as the ascent is shown to begin on the depth gauge. As the depth is shown or announced, have the students record the depth and the elapsed time on the stopwatch, which the teacher can announce as the movie is showing.
8. Using the depth and elapsed time data, have the students make a graph of the ascent according to the movie.
9. Have students discuss what the data from the movie tells them about the ascent of the submarine. Have them determine ways to calculate the ascent rate(s) of the submarine in the movie.

Discussion

1. How far does the submarine rise in this segment? How long did it take to rise that distance? (Answers should be similar to those in *Background Information.)*
2. How many meters does the submarine ascend each second? (Answers should be similar to those in *Background Information.)*
3. When you draw a graph that starts at –235 m of depth and goes up at the ascent rate you calculated, when does it reach the surface? (Answers should be similar to those in *Background Information.)*
4. How do you use numbers to determine how long it will take the submarine to reach the surface? [divide depth by rate of ascent]
5. From your graph and calculations, what range of time does the U-571 have from the start of its ascent until it reaches the surface? (Answers should be similar to those in *Background Information.)*

6. How does the graph or the movie's description of the submarine's ascent differ from the one you predicted off the depth gauge? [not straight, not as steep, the submarine rises slower]
7. What was the average ascent rate according to the movie? [235m/390s = 0.6 m/s]
8. By looking at the graph, when was the submarine rising slower than the average rate or faster then the average rate? [slower below –150 m, faster after –150 m.]
9. Why doesn't the calculated prediction match the actual movie? [willing suspension of disbelief, the movie was made for drama not accuracy]

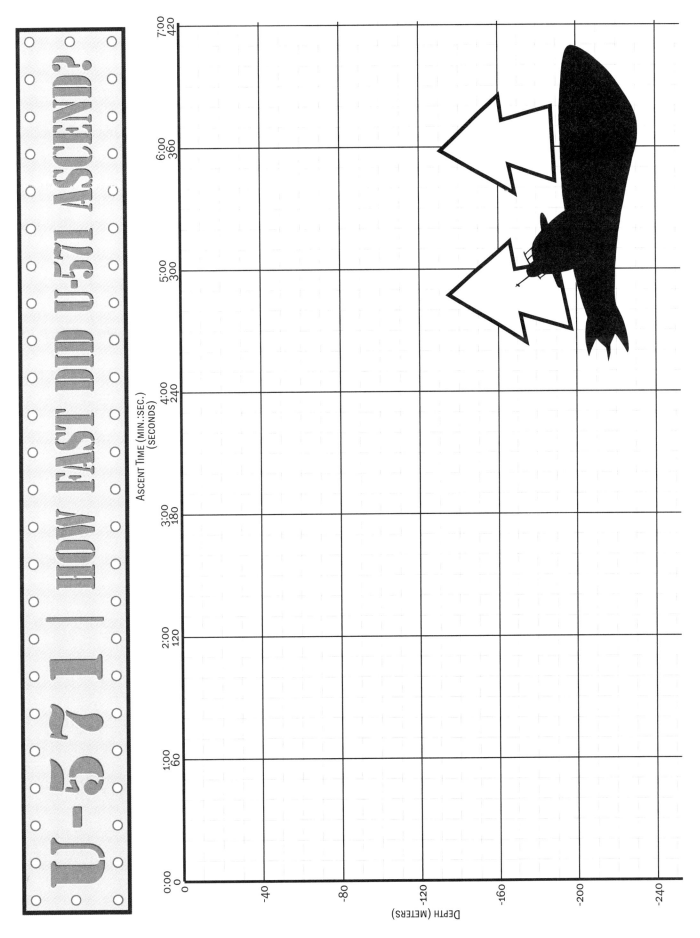

U-571 | HOW LONG DOES THE SUBMARINE HAVE UNTIL IT WILL SURFACE?

HOW LONG DOES THE SUBMARINE HAVE UNTIL IT WILL SURFACE?

DETERMINE FROM THE MOVIE HOW FAST THE SUBMARINE IS ASCENDING (ASCENT RATE).

	DEPTH AT START OF CLIP	DEPTH AT END OF CLIP	CHANGE IN DEPTH (M)	LENGTH OF CLIP (S)	ASCENT RATE (M/S)
FIRST CLIP					
SECOND CLIP					

USE WHAT YOU KNOW TO MAKE A GRAPH OF THE SUBMARINE'S ASCENT.

WRITE AN EQUATION THAT TELLS THE DEPTH OF THE SUBMARINE WHEN YOU KNOW HOW LONG IT HAS BEEN ASCENDING.

HOW LONG DOES IT TAKE THE SUBMARINE TO SURFACE IN THE MOVIE?

WATCH THE MOVIE AND START TIMING WHEN THE SUBMARINE BEGINS TO RISE. RECORD THE TIME ELAPSED AND THE DEPTHS SHOWN OR ANNOUNCED IN THE MOVIE UNTIL THE SUBMARINE SURFACES.

ELAPSED TIME										
DEPTH (METERS)										

PLOT THE DATA ON THE GRAPH TO SEE HOW THE SUBMARINE ASCENDED ACCORDING TO THE MOVIE.

- HOW DOES THE GRAPH OF THE SUBMARINE'S ASCENT THAT YOU CALCULATED DIFFER FROM THE GRAPH OF THE ASCENT ACCORDING TO THE MOVIE?

- HOW WOULD YOU EXPLAIN WHAT CAUSES THE DIFFERENCES?

OCTOBER SKY

October Sky is based on the true story of a NASA scientist. In 1957 in Coalwood, West Virginia, while most of the town mines the Earth, young Homer Hickman dreams of only the sky. Inspired by the dawn of the space age and the historic launching of the Soviet satellite Sputnik, Homer enlists his two best buddies as well as the school's resident math geek to help him design and launch space rockets. In order to be successful, Homer needs to learn a good deal of math and science. The movie portrays the effort Homer puts forth to learn what was necessary. When the law shuts down his effort to fly rockets, he applies his math skills to prove his innocence in the flight of a wayward rocket.

By looking at the data of rocket flights, students recognize patterns that allow them to develop an understanding of the basic formulas Homer used. In launching water or air rockets, students learn the methods of determining the height of a rocket's flight. By using the data in the film and their math knowledge, students prove the distance of Homer's rocket's flight and prove his innocence.

October Sky received a PG rating making it acceptable for most students. A clip containing a series of crashes (in the early part of the movie) is accompanied by only music. The scene where Homer explains the math he used to determine the height of the rocket has no objectionable language.

The book from which the movie was made, *October Sky: A Memoir,* provides interesting reading along with details and background not shown in the movie.

Topic
Patterns and functions

Key Question
How can you use the data of a falling rocket to predict how long it will take the rocket to hit the ground?

Learning Goals
Students will
1. recognize numeric and graphic patterns, and
2. describe those patterns using equations.

Guiding Documents
Project 2061 Benchmarks
- *Mathematicians often represent things with abstract ideas, such as numbers or perfectly straight lines, and then work with those ideas alone. The "things" from which they abstract can be ideas themselves (for example, a proposition about "all equal-sided triangles" or "all odd numbers").*
- *Mathematical statements can be used to describe how one quantity changes when another changes. Rates of change can be computed from magnitudes and vice versa.*
- *Graphs can show a variety of possible relationships between two variables. As one variable increases uniformly, the other may do one of the following: always keep the same proportion to the first, increase or decrease steadily, increase or decrease faster and faster, get closer and closer to some limiting value, reach some intermediate maximum or minimum, alternately increase and decrease indefinitely, increase and decrease in steps, or do something different from any of these.*

NRC Standards
- *The motion of an object can be described by its position, direction of motion, and speed. That motion can be measured and represented on a graph.*
- *An object that is not being subjected to a force will continue to move at a constant speed and in a straight line.*
- *Mathematics is important in all aspects of scientific inquiry.*

NCTM Standards 2000*
- *Represent, analyze, and generalize a variety of patterns with tables, graphs, words, and, when possible, symbolic rules*
- *Model and solve contextualized problems using various representations, such as graphs, tables, and equations*

Math
Patterns
Algebra
  graphing
  equations

Science
Physical science
  gravity
  acceleration

Integrated Processes
Observing
Comparing and contrasting
Collecting and organizing data
Predicting
Inferring
Interpreting data

Materials
Student pages

Background Information
After a rocket is launched and its fuel burned up, the most significant force acting on it is gravity. Gravity slows the rocket's vertical ascent until it stops. At that point, the rocket begins to fall and accelerates covering a greater distance each second. Scientists have found that objects in Earth's gravitational field accelerate at a rate of 32 ft./sec./sec. in a vacuum. The equation $S = 1/2at^2$ describes the distance an object has fallen. The S represents the distance fallen, a is the acceleration of gravity, and t is the time the object has fallen. When the acceleration of 32 ft./sec./sec. is substituted in the equation, it can be simplified to $S = 16t^2$.

Given time and height data of a falling rocket students can develop an equation after looking at the pattern between the time and the distance fallen. Using the data from the graph on the student page, they should get the following chart:

Time	Distance(d)
0	0
1	16
2	64
3	144
4	256
5	400

Students will find different patterns from which different, but equivalent, equations arise.

Some students will notice that all the distances are multiples of 16 and will find a pattern like this:

$$0 \bullet 16 = 0$$
$$1 \bullet 16 = 16$$
$$4 \bullet 16 = 64$$
$$9 \bullet 16 = 144$$
$$16 \bullet 16 = 256$$
$$25 \bullet 16 = 400$$

This shows that each 16 is multiplied by a perfect square number which has a square root equal to the time for that distance. The resulting equation is: $t \bullet t \bullet 16 = 16t^2 = d$

Other students may recognize the distances as square numbers.

Distance(d)	Sq. Root
0	0
16	4
64	8
144	12
256	16
400	20

As they develop a chart, these students will notice that the square root of each distance is four times larger than the time. They will suggest multiplying each time by four and then squaring the product. Their thinking can be symbolized: $(t \bullet 4) \bullet (t \bullet 4) = (4t)^2 = d$

Other students have suggested dividing the distance by the time.

$$16 \div 1 = 16$$
$$64 \div 2 = 32$$
$$144 \div 3 = 48$$
$$256 \div 4 = 64$$
$$400 \div 5 = 80$$

At this point most students recognize that the quotients are 16 times greater then the time. Using this reasoning, they would multiply the time by 16 to get the quotient and then multiply the quotient by the time to get the distance. Their thinking is symbolized: $(t \bullet 16) \bullet t = 16t^2 = d$.

Although all three of these equations are equivalent, it is a valuable experience to see how they are developed from very different thinking.

Many student will come to one of these ways after first considering finite differences. The first difference often bothers students because it is not constant; however, if they explore the difference of the differences, they will find a constant.

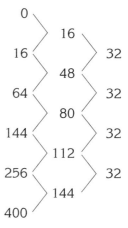

Although this method makes it difficult to generate an equation, it allows students to complete the chart and the graph. The chart brings up some very significant information. The first column is the total distance traveled. The first difference is the distance traveled each second or the average speed during each second. The second difference, the constant, is how much farther or faster the object is moving each second. It is the rate at which the speed of the object is accelerating. It is the acceleration constant of gravity on Earth, 32 ft./sec./sec.

In the movie *October Sky,* one way the "rocket boys" measure the height of the rocket is with fall time. In the movie they are accused of having one of their rockets start a fire. After a flight of a rocket like the one claimed to have started the fire, Quentin announces that the fall time was 12 seconds. Using the equation they have developed, students can determine the height of the rocket. $(d = 16t^2 = 16(12^2) = 2304)$

In another scene in the movie, Homer explains to the principal how their rocket could not have started the fire. He uses a fall time of 14 seconds and substitutes it in the generalized physics equation $S = 1/2\ at^2$. This provides an opportunity for students to check the accuracy of the film.

Management

1. Before class get a copy of the movie *October Sky* and identify the locations of the two scenes mentioned in *Background Information* to show to the class.
2. Students will need time to explore and search for patterns and then feel comfortable presenting their ideas to the rest of the class. A successful method is to ask students to work silently on their own looking for patterns for three to five minutes. Allow them to share with a partner what patterns they have found. After they have had some time to collaborate, ask for volunteers to share.

Procedure

1. Show the class the first segment (suggested in *Background Information*) of the movie *October Sky*. Have the students suggest how knowing the drop time might allow the "rocket boys" to determine the height of the rocket's flight.
2. Distribute the student page while discussing the *Key Question* with the students.
3. Using the data from the graph, have the students complete the chart through five seconds.
4. Allow the students time to find patterns that will allow them to complete the chart and make the graph.
5. Have the students share with the class the variety of patterns they found and how they used them to complete the chart.
6. Help students translate their patterns of solutions into equations.
7. Work as a class and show that the different equations are equivalent.
8. Have students use their equations to determine how high the rocket in the movie went. [2304 feet] Then watch the second scene from *October Sky* and have students determine the height of the rocket's flight (3136 feet) and decide whether the movie is accurate on the math and physics.

Discussion

1. What patterns did you find in the numbers that allowed you to complete the chart? (See *Background Information.*)
2. Explain how you used your pattern to complete the chart.
3. How could you translate the pattern you used to complete the chart into symbols? (See *Background Information.*)
4. How can you see the pattern of bigger and bigger numbers in the distance on the graph of the rocket? [bigger and bigger steps]
5. How might you be able to use your equation other than at looking at rockets? [determining the heights of cliffs and bluffs, measuring how high you throw a ball]

Extensions

1. Use your equation to determine how high you can throw a ball.
2. Watch one of the segments of *October Sky* described in *Background Information* and determine how high the "rocket boys'" rocket must have gone or check whether the math was done correctly in the movie.

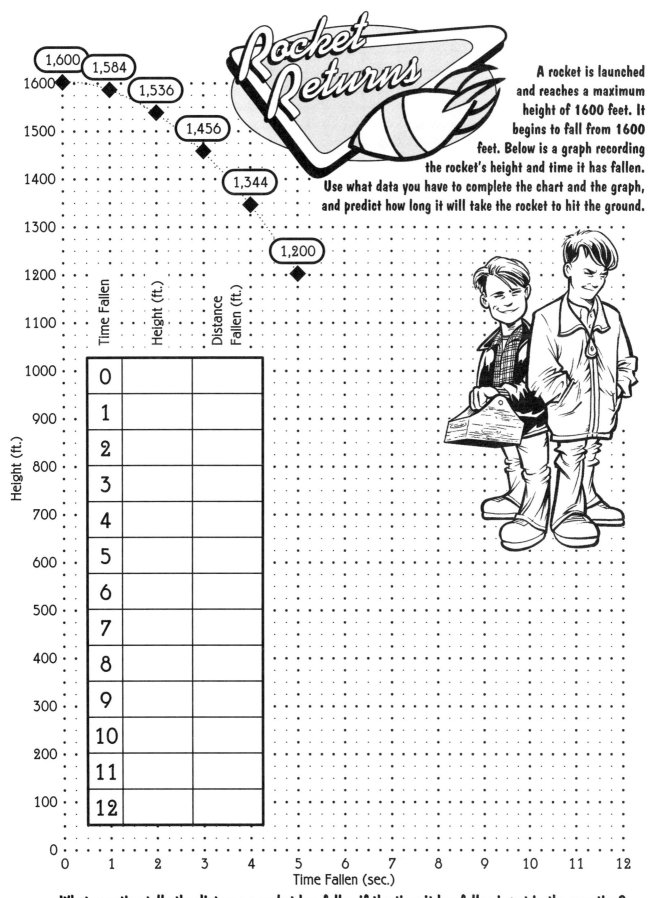

Rocket Returns

A rocket is launched and reaches a maximum height of 1600 feet. It begins to fall from 1600 feet. Below is a graph recording the rocket's height and time it has fallen. Use what data you have to complete the chart and the graph, and predict how long it will take the rocket to hit the ground.

1,600
1,584
1,536
1,456
1,344
1,200

Time Fallen	Height (ft.)	Distance Fallen (ft.)
0		
1		
2		
3		
4		
5		
6		
7		
8		
9		
10		
11		
12		

Height (ft.)

Time Fallen (sec.)

What equation tells the distance a rocket has fallen if the time it has fallen is put in the equation?

Topic
Patterns and functions

Key Question
If you know a stomp rocket is going to be shot into the air from a 112 foot high sand dune and has a vertical velocity of 96 feet per second, how could you predict its altitude for every second of its flight?

Learning Goals
Students will:
1. learn how the effect of gravity causes a parabolic flight for a rocket,
2. learn how each term in a polynomial describes an attribute of an object in free flight, and
3. recognize how each term in a polynomial is evident in its graph.

Guiding Documents
Project 2061 Benchmarks
- *Mathematicians often represent things with abstract ideas, such as numbers or perfectly straight lines, and then work with those ideas alone. The "things" from which they abstract can be ideas themselves (for example, a proposition about "all equal-sided triangles" or "all odd numbers").*
- *Mathematical statements can be used to describe how one quantity changes when another changes. Rates of change can be computed from magnitudes and vice versa.*
- *Graphs can show a variety of possible relationships between two variables. As one variable increases uniformly, the other may do one of the following: always keep the same proportion to the first, increase or decrease steadily, increase or decrease faster and faster, get closer and closer to some limiting value, reach some intermediate maximum or minimum, alternately increase and decrease indefinitely, increase and decrease in steps, or do something different from any of these.*

NRC Standards
- *The motion of an object can be described by its position, direction of motion, and speed. That motion can be measured and represented on a graph.*
- *An object that is not being subjected to a force will continue to move at a constant speed and in a straight line.*
- *Mathematics is important in all aspects of scientific inquiry.*

NCTM Standards 2000*
- *Represent, analyze, and generalize a variety of patterns with tables, graphs, words, and, when possible, symbolic rules*
- *Relate and compare different forms of representation for a relationship*
- *Identify functions as linear or nonlinear and contrast their properties from tables, graphs, or equations*
- *Model and solve contextualized problems using various representations, such as graphs, tables, and equations*

Math
Patterns
Algebra
    graphing
    equations
    linear functions
    quadratic functions

Science
Physical science
    gravity
    acceleration
    rocket flight

Integrated Processes
Observing
Collecting and organizing data
Predicting
Inferring
Interpreting data

Materials
Student pages
Scissors
Small binder clips

Background Information

The altitude of an air-driven projectile, like a stomp rocket or bullet, at any given point in its flight depends on several factors including the initial launch altitude, the initial velocity, and the acceleration of gravity.

Consider the problem suggested in this investigation: A stomp rocket is going to be shot into the air from a 112-foot high sand dune and have a vertical velocity of 96 feet per second. The 112-foot high dune is the initial altitude to which other changes will need to be added. If students have studied rates and speeds, they will be quick to add the height gained with the 96 feet per second. They will build a chart like this one to show their predictions:

Time (seconds):	0	1	2	3	4	5	6
Predicted Height:	112	208	304	400	496	592	688

Each height is an increase of 96 feet from the previous height. This results in a graph with a straight line with a starting point (y-intercept) of 112 and a slope of 96. This pattern can be represented in the equation: $H = 112 + 96t$.

As students gather data from a realistic animation of the flight, the data contradict their predictions. The rocket flies in a parabolic curve rather than the straight line of consistently increasing height. Likewise, the graph of the data produces a parabola rather than the predicted line. As students study the difference in their prediction and the height in the movie, interesting patterns emerge.

Time (seconds):	0	1	2	3	4	5	6	7
Predicted Height:	112	208	304	400	496	592	688	784
Movie Height:	112	192	240	256	240	192	112	0
Difference:	0	16	64	144	256	400	576	784

First students recognize a numeric symmetry in the movie height where it is increasing and decreasing with the same numbers. This symmetrical patterned is mirrored in the graph's parabola formed by the data. This matches our intuitive experience that what goes up must come down, rather then the forever rising prediction.

The second pattern to emerge is the ever-increasing amount of difference in the prediction and movie heights. If student have done the preceding investigation, *Rocket Returns*, they will recognize these differences are caused by the acceleration of gravity. These numbers can be represented with the term $16t^2$. This difference can be seen graphically by the ever-increasing gap between the line of the predicted height and the parabolic curve of the movie height. The predicted line would be correct if it was moved down the difference at each time. Symbolically we represented the line with the equation: $H = 112 + 96t$. Now we take away the difference and get the corrected equation: $H = 112 + 96t - 16t^2$. If we write this in the accepted mathematical convention of decreasing powers we get the equation: $H = -16t^2 + 96t + 112$. Each constant in this equation informs us about the situation. The coefficient -16 speaks to the effects of gravity on Earth. Ninety-six is the initial velocity and 112 is the initial height. All projectiles in Earth's gravitational field will have similar equations except the 96 and 112 will be substituted with numbers relevant to the situation.

The actual "rocket boys" portrayed in the movie *October Sky* assumed for their calculations that all the rocket fuel was expended by the time the rocket left the launch pad. This is not precisely true for the rockets they flew or for water rockets that might be launched in class, but it does provide an estimation since the fuel is spent so rapidly after launch.

Using the general equation developed in this investigation students could use the information they gathered from the movie *October Sky* to determine the initial speed of their AUK XII rocket. Since it had a fall time of 14 seconds they can use the equation $H = 16t^2$ to determine it went 3136 feet up. Now this information along with the time can be substituted into an equation like the one developed in the investigation. (v represents the unknown initial vertical velocity.) Working backwards, a solution for v is found.

$$
\begin{aligned}
3136 &= -16(14^2) + v(14) + 0 \\
3136 &= -3136 + 14v \\
6272 &= 14v \\
448 &= v
\end{aligned}
$$

The rocket boys would have estimated their rocket reached an initial velocity of 448 feet per second (a little more then 300 mph).

Management

1. Before class make at least one flip book following these instructions:
   a. Cut out all 36 frames on the lines.
   b. Stack the frames face up in order with the one with *time = 0* on top.
   c. Align all the frames edge to edge before securing them as a pad with a small binder clip on the left edge.
   d. Holding the clip in your left hand and setting the right edge of the frames behind your right thumb, slowly rotate your left wrist so the frames are pulled past your right thumb. You will see an animation of rocket flight.

Note: You may find the flip book works better if each frame is slightly to the right of the one above it. It also helps to "work" in the book a while to get the animation to flow smoothly.

2. You may allow every student to make a flip book. This tends to be quite time consuming and uses a good deal of paper. It is more efficient to have each group of four students work together to make one flip book.

3. A page with the frame from every whole second is provided if a teacher wants to use only the data. The sample flip book may be used for students to verify that the animation simulates rocket flight.

Procedure

1. Distribute the student page and ask the students to discuss the *Key Question*.
2. Have students predict, record, and graph the height of a rocket that starts at 112 feet and goes up 96 feet every second.
3. Develop with students' help the equation that would represent the height of the rocket at any time according to their prediction.
4. Provide the students with materials to construct a flip book animation of the rocket flight or distribute the page with the eight whole second frames.
5. Using the flip book for data, have students record and graph the height of the rocket each second and calculate the difference between their predictions and the animation's heights.
6. Have students discuss what patterns they see in their chart and graph.
7. Encourage students to write an algebraic expression to represent the pattern in the difference column and ask them how this difference is represented on the graph.
8. Ask the class to modify the equation for the students' predicted height to take into account the difference from the animated height. Ask students what each of the constants tells them about the situation.

Discussion

1. What equation tells the predicted flight of the rocket at any given time? [$H = 112 + 96t$]
2. What do each of the constants in the equation represent? [launch altitude, vertical speed or velocity]

3. How does each of the constants in the equation show up in the graph? [y-intercept, slope or rise in line each second]
4. What patterns do you see in height of the rocket in the animation? [The rocket goes up and then comes down. The height numbers are the same on either side of three seconds, 256 feet. The numbers are symmetrical around 256.]
5. How is the graph of this data similar to the number patterns? [goes up then down, same position on either side of three seconds, symmetrical around three seconds]
6. What pattern do you see in the differences of your prediction and the animation? [The differences get larger each time, the same as the gravity pattern.]
7. What symbolic expression would describe this pattern? [$16t^2$]
8. Where does this difference show up on the graph? [space between predicted line and animated curve]
9. What do you have to do to the predicted height to get the animated height? [subtract the difference]
10. How can you modify the equation for the prediction to show this difference? [$H = 112 + 96t - 16t^2$]
11. What does each of the constants in the equation tell you about the situation? [initial height, initial velocity, gravity]
12. How could you use this equation or one like it?

Extensions

1. Have students modify their equation and work backwards to solve for the velocity of the AUK XII rocket in *October Sky* that had a fall time of 14 seconds.
2. Have students determine the heights of balls they throw up into the air. Using their data and equation, have them calculate the ball's velocity.

* Reprinted with permission from *Principles and Standards for School Mathematics,* 2000 by the National Council of Teachers of Mathematics. All rights reserved.

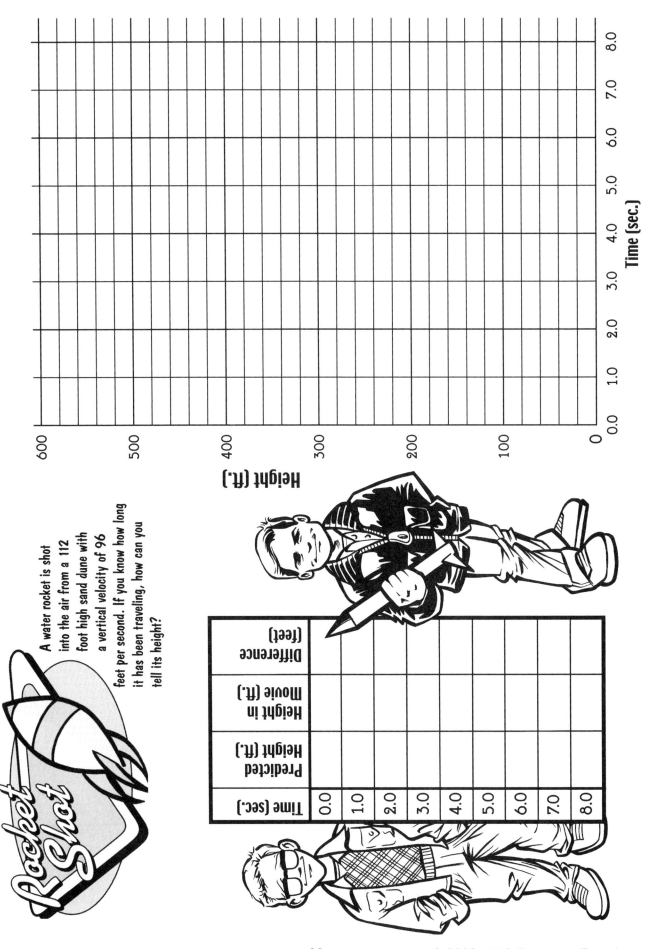

Rocket Shot

A water rocket is shot into the air from a 112 foot high sand dune with a vertical velocity of 96 feet per second. If you know how long it has been traveling, how can you tell its height?

Time (sec.)	Predicted Height (ft.)	Height in Movie (ft.)	Difference (feet)
0.0			
1.0			
2.0			
3.0			
4.0			
5.0			
6.0			
7.0			
8.0			

Height (ft.)

Time (sec.)

Panel	TIME	ALTITUDE
1	0.0	112
2	0.2	131
3	0.4	149
4	0.6	164
5	0.8	179
6	1.0	192
7	1.2	204
8	1.4	215
9	1.6	225

Each panel has a vertical altitude scale labeled: 0, 50, 100, 150, 200, 250

	1.8 TIME		2.0 TIME		2.2 TIME
250	**233** ALTITUDE	250	**240** ALTITUDE	250	**246** ALTITUDE
200		200		200	
150		150		150	
100		100		100	
50		50		50	
0		0		0	

	2.4 TIME		2.6 TIME		2.8 TIME
250	**250** ALTITUDE	250	**253** ALTITUDE	250	**255** ALTITUDE
200		200		200	
150		150		150	
100		100		100	
50		50		50	
0		0		0	

	3.0 TIME		3.2 TIME		3.4 TIME
250	**256** ALTITUDE	250	**255** ALTITUDE	250	**253** ALTITUDE
200		200		200	
150		150		150	
100		100		100	
50		50		50	
0		0		0	

3.6 TIME	3.8 TIME	4.0 TIME
250 —	250 —	250 —
250 ALTITUDE	**246** ALTITUDE	**240** ALTITUDE
200 —	200 —	200 —
150 —	150 —	150 —
100 —	100 —	100 —
50 —	50 —	50 —
0 —	0 —	0 —

4.2 TIME	4.4 TIME	4.6 TIME
250 —	250 —	250 —
233 ALTITUDE	**225** ALTITUDE	**215** ALTITUDE
200 —	200 —	200 —
150 —	150 —	150 —
100 —	100 —	100 —
50 —	50 —	50 —
0 —	0 —	0 —

4.8 TIME	5.0 TIME	5.2 TIME
250 —	250 —	250 —
204 ALTITUDE	**192** ALTITUDE	**179** ALTITUDE
200 —	200 —	200 —
150 —	150 —	150 —
100 —	100 —	100 —
50 —	50 —	50 —
0 —	0 —	0 —

5.4 TIME / **164** ALTITUDE	**5.6** TIME / **148** ALTITUDE	**5.8** TIME / **131** ALTITUDE
6.0 TIME / **112** ALTITUDE	**6.2** TIME / **92** ALTITUDE	**6.4** TIME / **71** ALTITUDE
6.6 TIME / **49** ALTITUDE	**6.8** TIME / **25** ALTITUDE	**7.0** TIME / **0** ALTITUDE

Sample Flip Book

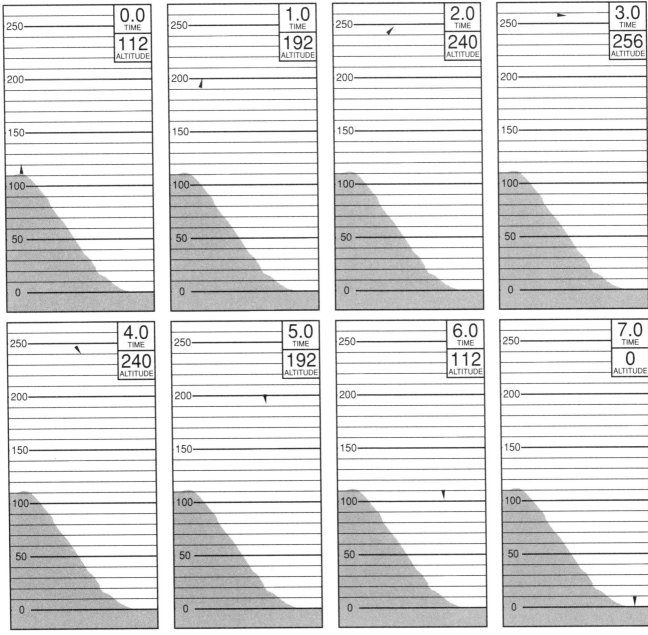

Topic
Proportions and Similarity

Key Question
How can you measure the height of a rocket from observing its flight?

Learning Goals
Students will:
1. make a scale drawing of a rocket flight using appropriate tools for angular measure, and
2. use proportions to make indirect measurements from scale drawings.

Guiding Documents
Project 2061 Benchmarks
- *Mathematics is helpful in almost every kind of human endeavor—from laying bricks to prescribing medicine or drawing a face. In particular, mathematics has contributed to progress in science and technology for thousands of years and still continues to do so.*
- *Estimate distances and travel times from maps and the actual size of objects from scale drawings.*
- *Use calculators to compare amounts proportionally.*

NRC Standards
- *Use appropriate tools and techniques to gather, analyze, and interpret data.*
- *The motion of an object can be described by its position, direction of motion, and speed. That motion can be measured and represented on a graph.*

NCTM Standards 2000*
- *Develop, analyze, and explain methods for solving problems involving proportions, such as scaling and finding equivalent ratios*
- *Solve problems involving scale factors, using ratio and proportion*
- *Understand, select, and use units of appropriate size and type to measure angles, perimeter, area, surface area, and volume*
- *Recognize and apply geometric ideas and relationships in areas outside the mathematics classroom, such as art, science, and everyday life*

Math
Proportional reasoning
  scaling
Geometry
  similarity
Measurement
  angle
  indirect measurement

Science
Physical science
  rocket flight

Integrated Processes
Observing
Comparing and contrasting
Collecting and organizing data
Predicting
Inferring
Interpreting data

Materials
Rocket (water or stomp)
30 m tape measure
Card stock
Scissors
Tape
Straws
String
Paper clips
Rulers
Pennies

Background Information
Water rockets and air-driven rockets (stomp rockets) provide a convenient and inexpensive way to study rocketry. A typical question to arise concerns the maximum altitude of the rocket. This question can be answered by its fall time, although this is an approximation because the formula used does not account for air resistance. The most accurate way, and the method used by scientists, is triangulation.

Tracking a rocket in flight forms a triangle with the sides being the lines between the rocket and the tracking position one, the rocket and tracking position two, and the distance between the two tracking positions. Using a scale drawing of this triangle, or trigonometry, the height can be determined.

Using the two angles of inclination from the tracking positions and the scaled distance between the two tracking positions can make a similar triangle to the actual one formed by the launch. A water rocket or air rocket is launched vertically. This allows one of the tracking positions to be replaced by the launch site. The angle of the rocket is 90° relative to the ground at the launch site. The distance from the launch site to a tracking position is easily measured on the ground. Using a theodolite, a plumb line attached to a protractor, the students can measure the angle of inclination from the tracking position of the rocket when it reaches its maximum height. Knowing the two angles and the side between them allows one to construct a similar triangle using the angle-side-angle theorem of geometry.

Students can make a scale drawing of the rocket flight by first drawing two lines at right angles to each other. The vertical line represents the rocket's ascent. The horizontal line represents the ground. A point on the horizontal line can be chosen to represent the tracking position. Using a protractor to measure the angle of inclination, a line of sight can be drawn between the tracking position on the horizontal line and the vertical line. The intersection of this diagonal line and the vertical line represents the position of the rocket at the maximum height.

The length of the vertical line on the drawing from the horizontal line to the point of intersection represents the maximum height of the rocket's flight. When this segment is measured, it can be used in the proportion to solve the actual height of the rocket. The proportion below shows how it might be set up if a scale were used.

Horizontal	Vertical
Distance between tracking and launch	Distance between ground and rocket

$$\frac{\text{Length of Segment on Drawing}}{\text{Actual Measured Distance}} = \frac{\text{Length of Segment on Drawing}}{\text{Calculate Unknown Height}}$$

Students might choose another proportion to solve the problem that is related to trigonometry. The ratio made by the measurements of the drawing is the trigonometric tangent ratio for the angle of inclination of the flight. If students choose this proportion, it might provide a natural opportunity to introduce the concept of trigonometry and its application. A tangent table is included should such an opportunity arise.

Drawing	Actual
Length of Vertical Segment	Calculate Vertical Segment

$$\frac{\text{Length of Vertical Segment}}{\text{Length of Horizontal Segment}} = \frac{\text{Calculate Vertical Segment}}{\text{Length of Horizontal Segment}}$$

In reality, the rockets do not always fly straight up. Placing students as trackers relatively evenly spaced around a circle with the launch sight as center can compensate for this. Each tracker needs to be the same distance from the launch sight. Finding the mean of the angle of inclination from all the trackers, will provide an average that is more accurate.

This method is based on tracking the rocket from eye height and the calculated height is the height of the rocket above the eye. To get an accurate measurement, the eye height must be added to the calculated height. This can be an important factor on the relatively low flights of these types of rockets.

In the movie *October Sky,* this method of tracking was only alluded to as Quentin and Homer mention trigonometry several times in the discussions. However, in the book, Homer Hickman clearly states that triangulation and trigonometry were used. The selection from the book is included as a student page to be read by the class and the method discussed and tried in this investigation. The application of the trigonometric ratios is not included in this activity but may be included if it is appropriate for students.

Management

1. Each group of trackers will need a theodolite. These are best constructed by the teacher before class following these instructions.
   a. Copy the theodolites onto card stock.
   b. Cut out the theodolites along the solid line border.
   c. Using a pushpin or the end of a paper clip, make a hole in the centering point of the protractor where the 0° and 90° lines intersect.
   d. Thread a length of string through the hole so that about an inch of string goes through to the back and secure the string with tape on the back of the theodolite.
   e. Cut the string on the front so that at least two inches extends past the protractor's scale.
   f. Tie a paper clip to the free end of the string to act as a plumb bob. A penny may be secured in the clip to provide more mass.
   g. Tape a straw to the front of the theodolite as indicated.
2. Before the activity, students will need to be instructed in using the theodolite for tracking.
   a. Have one student from each group track (follow) the flight of the rocket through the straw. Students should look through the 90° end of the straw, holding the theodolite in such a way that the plumb bob and line are free to move.
   b. When the rocket reaches its maximum height, the tracker stops tracking the rocket and holds the theodolite in this final position.
   c. A second student notes the position of where the string crosses the protractor's scale as the angle of inclination.

3. Before doing this activity, the teacher should be experienced in rocket launch. Prepare by going out to an appropriate field that is level, clear overhead, and has space to spread the trackers on all sides of the launch. The teacher should practice launching the rocket as vertically as possible. This will provide an opportunity to verify that the field is large enough for launching and tracking.

4. For the most accurate measurement, numerous teams of trackers evenly surrounding the launch site works best. Each tracking team can be made of four students: a tracker, the angle reader, and two members to use a tape measure to determine the distance from the launch site to the tracking position.

Procedure

1. Have the students read the selection from *October Sky* and then have them discuss the *Key Question* or how they might modify the method described in the reading to determine the height of their rocket.

2. Distribute the measuring tools to each group and instruct the students on how to use the theodolite (see *Management 2*). Make sure the students have developed a set of steps similar to the ones listed in this procedure, or clarify to the class what the steps are and what data they need to gather about the rocket's flight.

3. As a class go to the chosen location and evenly spread the tracking groups around the chosen launch site.

4. After the launch location has been marked, direct each group to measure and record the distance to their tracking position from the launch site. If you have chosen to average the angles of inclination as suggested in *Background Information,* instruct the groups to be the same designated distance from the launch site.

5. When all the groups are prepared, launch the rocket while the students track it, and measure and record the angle of inclination at the rocket's maximum height. It is suggested that several practice launches be done to allow students practice in tracking before the official flight takes place.

6. Invite each group to share the angle of inclination with the class if the class is using the averaging method. Have the students then calculate the average angle of inclination.

7. Direct each student to measure and draw a line with the angle of inclination on the record sheet.

8. After measuring the horizontal and vertical line segments on the drawing, have students develop a proportional relationship from the measurements of the drawing and the actual distance the tracker was from the launch site. (See *Background Information.*) Using proportional reasoning, the students calculate the height of the rocket's flight.

9. Have the students share the ways they set up the proportion and solved for the height of the rocket. Emphasize the variety of proportional relationships and their representations that arise from the data. This discussion should lead to the understanding that the eye height of the tracker needs to be added to the calculated height. If students suggest the appropriate proportional relationship, the tangent ratio might be discussed at this point.

10. Focus the students' attention back on the excerpt from *October Sky,* and have them identify how what they did was similar to what Quentin did.

Discussion

1. If the rocket did not go straight up, how did it show up in the class's data? [different angles of inclination]

2. How did averaging adjust for this error? [When one person is too high, the person directly across from him or her would be too low. Their average would adjust for the high-low situation.]

3. What does the horizontal side of the triangle on the drawing represent? [the distance between launch and tracker]

4. What does the vertical side of the triangle on the drawing represent? [the rocket's flight to its maximum height]

5. How is the drawing like the rocket's flight at the maximum height? [It is a scaled drawing of the flight. They're similar shapes so they share the same proportions.]

6. What proportional relationship did you see that let you determine the height of the rocket's flight? (See *Background Information.*)

7. How did you use the proportional relationship to calculate the height of the rocket? (See *Background Information.*)

8. The tracking was made at eye height. How does this affect your calculated height for the rocket? [add in eye height]

9. How could you change the measurement and calculation procedures to make the process more efficient? [put scaled measures on vertical line so height can be read directly from scale on vertical line, use trigonometry]

10. Compare what you did to determine the height of the rocket with what Quentin did in *October Sky*. [measure angle with theodolite, measured distance (tape measure vs. pacing), scaled drawing vs. trigonometric ratios]

11. How is what you did similar to what NASA does as it tracks rockets with radar? [trios of radar measuring angles of incidence along with known distances between tracking stations allows computers to use the proportional relationships of trigonometric ratios to calculate altitude]

12. How could you use this process to measure other objects?

Extensions

1. If the class has developed a more efficient system or trigonometry has been introduced, go out and launch the rocket again. Use the new method to determine the height of the rocket.

2. Have student use the triangulation method and the fall time method (see *Rocket Returns*) to see how similar the calculated heights are.

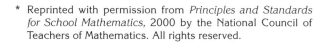

How can you modify the method the "rocket boys" used to determine the height of the rocket's flight?

This was also our first launch with an electrical-ignition system. I touched a wire to a car battery (an old one O'Dell got for free from a War junkyard), and Auk XII shot off the pad and leaned downrange. Quentin ran outside the bunker and fumbled with a new invention he called a "theodolite." It was a broomstick with an upside down protractor attached on one end and a wooden straightedge on the opposing side that rotated around a nail. He jammed the stick in the slack and went down on his knees and squinted along the straightedge at the rising rocket, smoke squirming from its tail, white against the brilliant blue of the cloudless sky. At the height of the Auk's climb, Quentin looked at the angle the ruler made with the protractor and called it out, then snatched a pencil stuck behind his ear and wrote it down on a scrap of paper. If his theodolite worked, trigonometry would give us the altitude of our rocket.

Auk XII's exhaust trail was still a fast stream when the rocket faltered and began to fall. It continued to smoke vigorously even after it struck the slack. While our audience cheered, we ran after our rocket and watched the last of its sputtering rocket candy burn up. I immediately saw the reason our rocket had lost its thrust. "The nozzle's gone," I told the others. "It must have blown out."

We looked closer. The weld was intact. The center of the nozzle was simply eaten away. Quentin came stepping up to us. "Three hundred and forty-eight," he said, finishing his count by bringing both his feet together at the final step. "I'm figuring about two point seven-five feet per step. That would be"—he made a quick mental calculation—"nine hundred and fifty-seven feet." Jake's trig book was under his arm. He ran his finger down the functions in the back. "Let's see, the tangent of forty degrees is about point eight-four. Call it point eight. Multiply that by nine hundred and sixty . . ."

We waited anxiously while Quentin worked it out in his head. It didn't take long. "Seven hundred and sixty feet!"

Reprinted with permission from: Hickman, Homer. *October Sky: A Memoir.* Dell Books, 1999. pp. 190-191 (ISBN: 0440235502)

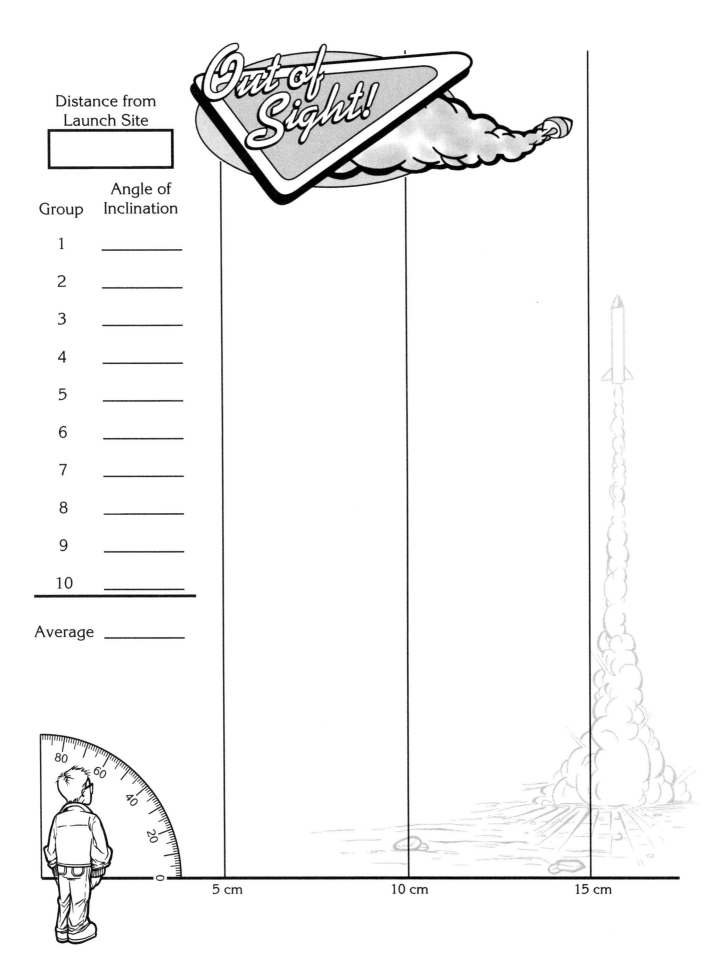

Distance from
Launch Site

Group	Angle of Inclination
1	_____
2	_____
3	_____
4	_____
5	_____
6	_____
7	_____
8	_____
9	_____
10	_____

Average _____

5 cm   10 cm   15 cm

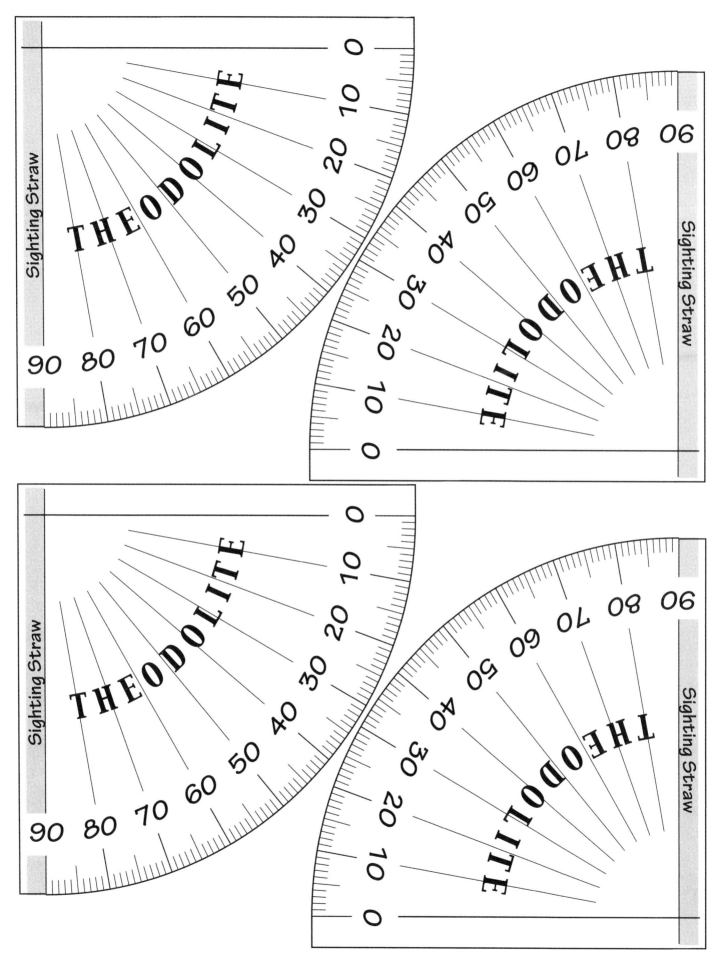

Angle	Tangent	Angle	Tangent	Angle	Tangent
1	.0175	31	.6009	61	1.804
2	.0349	32	.6249	62	1.881
3	.0524	33	.6494	63	1.963
4	.0699	34	.6745	64	2.050
5	.0875	35	.7002	65	2.145
6	.1051	36	.7265	66	2.246
7	.1228	37	.7536	67	2.356
8	.1405	38	.7813	68	2.475
9	.1584	39	.8098	69	2.605
10	.1763	40	.8391	70	2.747
11	.1944	41	.8693	71	2.904
12	.2126	42	.9004	72	3.078
13	.2309	43	.9325	73	3.271
14	.2493	44	.9657	74	3.487
15	.2679	45	1.000	75	3.732
16	.2867	46	1.035	76	4.011
17	.3057	47	1.072	77	4.331
18	.3249	48	1.111	78	4.705
19	.3443	49	1.150	79	5.145
20	.3640	50	1.192	80	5.671
21	.3839	51	1.235	81	6.314
22	.4040	52	1.280	82	7.115
23	.4245	53	1.327	83	8.144
24	.4452	54	1.376	84	9.514
25	.4663	55	1.428	85	11.43
26	.4877	56	1.483	86	14.30
27	.5095	57	1.540	87	19.08
28	.5317	58	1.600	88	28.64
29	.5543	59	1.664	89	57.29
30	.5774	60	1.732	90	∞

The AIMS Program

AIMS is the acronym for "**A**ctivities **I**ntegrating **M**athematics and **S**cience." Such integration enriches learning and makes it meaningful and holistic. AIMS began as a project of Fresno Pacific University to integrate the study of mathematics and science in grades K-9, but has since expanded to include language arts, social studies, and other disciplines.

AIMS is a continuing program of the non-profit AIMS Education Foundation. It had its inception in a National Science Foundation funded program whose purpose was to explore the effectiveness of integrating mathematics and science. The project directors in cooperation with 80 elementary classroom teachers devoted two years to a thorough field-testing of the results and implications of integration.

The approach met with such positive results that the decision was made to launch a program to create instructional materials incorporating this concept. Despite the fact that thoughtful educators have long recommended an integrative approach, very little appropriate material was available in 1981 when the project began. A series of writing projects have ensued, and today the AIMS Education Foundation is committed to continue the creation of new integrated activities on a permanent basis.

The AIMS program is funded through the sale of books, products, and staff development workshops and through proceeds from the Foundation's endowment. All net income from program and products flows into a trust fund administered by the AIMS Education Foundation. Use of these funds is restricted to support of research, development, and publication of new materials. Writers donate all their rights to the Foundation to support its on-going program. No royalties are paid to the writers.

The rationale for integration lies in the fact that science, mathematics, language arts, social studies, etc., are integrally interwoven in the real world from which it follows that they should be similarly treated in the classroom where we are preparing students to live in that world. Teachers who use the AIMS program give enthusiastic endorsement to the effectiveness of this approach.

Science encompasses the art of questioning, investigating, hypothesizing, discovering, and communicating. Mathematics is the language that provides clarity, objectivity, and understanding. The language arts provide us powerful tools of communication. Many of the major contemporary societal issues stem from advancements in science and must be studied in the context of the social sciences. Therefore, it is timely that all of us take seriously a more holistic mode of educating our students. This goal motivates all who are associated with the AIMS Program. We invite you to join us in this effort.

Meaningful integration of knowledge is a major recommendation coming from the nation's professional science and mathematics associations. The American Association for the Advancement of Science in *Science for All Americans* strongly recommends the integration of mathematics, science, and technology. The National Council of Teachers of Mathematics places strong emphasis on applications of mathematics such as are found in science investigations. AIMS is fully aligned with these recommendations.

Extensive field testing of AIMS investigations confirms these beneficial results:

1. Mathematics becomes more meaningful, hence more useful, when it is applied to situations that interest students.
2. The extent to which science is studied and understood is increased, with a significant economy of time, when mathematics and science are integrated.
3. There is improved quality of learning and retention, supporting the thesis that learning which is meaningful and relevant is more effective.
4. Motivation and involvement are increased dramatically as students investigate real-world situations and participate actively in the process.

We invite you to become part of this classroom teacher movement by using an integrated approach to learning and sharing any suggestions you may have. The AIMS Program welcomes you!

AIMS Education Foundation Programs

Practical proven strategies to improve student achievement

When you host an AIMS workshop for elementary and middle school educators, you will know your teachers are receiving effective usable training they can apply in their classrooms immediately.

Designed for teachers—AIMS Workshops:
- Correlate to your state standards;
- Address key topic areas, including math content, science content, problem solving, and process skills;
- Teach you how to use AIMS' effective hands-on approach;
- Provide practice of activity-based teaching;
- Address classroom management issues, higher-order thinking skills, and materials;
- Give you AIMS resources; and
- Offer college (graduate-level) credits for many courses.

Aligned to district and administrator needs—AIMS workshops offer:
- Flexible scheduling and grade span options;
- Custom (one-, two-, or three-day) workshops to meet specific schedule, topic and grade-span needs;
- Pre-packaged one-day workshops on most major topics—only $3,900 for up to 30 participants (includes all materials and expenses);
- Prepackaged *week-long* workshops (four- or five-day formats) for in-depth math and science training—only $12,300 for up to 30 participants (includes all materials and expenses);
- Sustained staff development, by scheduling workshops throughout the school year and including follow-up and assessment;
- Eligibility for funding under the Eisenhower Act and No Child Left Behind; and
- Affordable professional development—save when you schedule consecutive-day workshops.

University Credit—Correspondence Courses

AIMS offers correspondence courses through a partnership with Fresno Pacific University.
- Convenient distance-learning courses—you study at your own pace and schedule. No computer or Internet access required!

The tuition for each three-semester unit graduate-level course is $264 plus a materials fee.

The AIMS Instructional Leadership Program

This is an AIMS staff-development program seeking to prepare facilitators for leadership roles in science/math education in their home districts or regions. Upon successful completion of the program, trained facilitators become members of the AIMS Instructional Leadership Network, qualified to conduct AIMS workshops, teach AIMS in-service courses for college credit, and serve as AIMS consultants. Intensive training is provided in mathematics, science, process and thinking skills, workshop management, and other relevant topics.

Introducing AIMS Science Core Curriculum

Developed in alignment with your state standards, AIMS' Science Core Curriculum gives students the opportunity to build content knowledge, thinking skills, and fundamental science processes.
- *Each* grade specific module has been developed to extend the AIMS approach to full-year science programs.
- *Each* standards-based module includes math, reading, hands-on investigations, and assessments.

Like all AIMS resources these core modules are able to serve students at all stages of readiness, making these a great value across the grades served in your school.

For current information regarding the programs described above, please complete the following:

Information Request

Please send current information on the items checked:

_____ *Basic Information Packet* on AIMS materials   ____ Hosting information for AIMS workshops
_____ *AIMS Instructional Leadership Program*   ____ AIMS Science Core Curriculum

Name _____   Phone _____

Address_____
        Street   City   State   Zip

Magazine

**YOUR K-9 MATH AND SCIENCE
CLASSROOM ACTIVITIES RESOURCE**

The AIMS Magazine is your source for standards-based, hands-on math and science investigations. Each issue is filled with teacher-friendly, ready-to-use activities that engage students in meaningful learning.

• *Four issues each year (fall, winter, spring, and summer).*

Current issue is shipped with all past issues within that volume.

1821	Volume	XXI	2006-2007	$19.95
1822	Volume	XXII	2007-2008	$19.95

Two-Volume Combination

M20507	Volumes XX & XXI	2005-2007	$34.95
M20608	Volumes XXI & XXII	2006-2008	$34.95

Back Volumes Available
Complete volumes available for purchase:

1802	Volume II	1987-1988	$19.95
1804	Volume IV	1989-1990	$19.95
1805	Volume V	1990-1991	$19.95
1807	Volume VII	1992-1993	$19.95
1808	Volume VIII	1993-1994	$19.95
1809	Volume IX	1994-1995	$19.95
1810	Volume X	1995-1996	$19.95
1811	Volume XI	1996-1997	$19.95
1812	Volume XII	1997-1998	$19.95
1813	Volume XIII	1998-1999	$19.95
1814	Volume XIV	1999-2000	$19.95
1815	Volume XV	2000-2001	$19.95
1816	Volume XVI	2001-2002	$19.95
1817	Volume XVII	2002-2003	$19.95
1818	Volume XVIII	2003-2004	$19.95
1819	Volume XIX	2004-2005	$19.95
1820	Volume XX	2005-2006	$19.95

Volumes II to XIX include 10 issues.

Call 1.888.733.2467 or go to www.aimsedu.org

Subscribe
to the
AIMS Magazine

$19.95
a year!

AIMS Magazine
is published four
times a year.

Subscriptions ordered
at any time will receive
all the issues for that
year.

AIMS Online—www.aimsedu.org

To see all that AIMS has to offer, check us out on the Internet at www.aimsedu.org. At our website you can search our activities database; preview and purchase individual AIMS activities; learn about core curriculum, college courses, and workshops; buy manipulatives and other classroom resources; and download free resources including articles, puzzles, and sample AIMS activities.

AIMS News
While visiting the AIMS website, sign up for AIMS News, our FREE e-mail newsletter. You'll get the latest information on what's new at AIMS including:

• New publications;
• New core curriculum modules; and
• New materials.

Sign up today!

AIMS Program Publications

Actions with Fractions, 4-9
Awesome Addition and Super Subtraction, 2-3
Bats Incredible! 2-4
Brick Layers II, 4-9
Chemistry Matters, 4-7
Counting on Coins, K-2
Cycles of Knowing and Growing, 1-3
Crazy about Cotton, 3-7
Critters, 2-5
Electrical Connections, 4-9
Exploring Environments, K-6
Fabulous Fractions, 3-6
Fall into Math and Science, K-1
Field Detectives, 3-6
Finding Your Bearings, 4-9
Floaters and Sinkers, 5-9
From Head to Toe, 5-9
Fun with Foods, 5-9
Glide into Winter with Math and Science, K-1
Gravity Rules! 5-12
Hardhatting in a Geo-World, 3-5
It's About Time, K-2
It Must Be A Bird, Pre-K-2
Jaw Breakers and Heart Thumpers, 3-5
Looking at Geometry, 6-9
Looking at Lines, 6-9
Machine Shop, 5-9
Magnificent Microworld Adventures, 5-9
Marvelous Multiplication and Dazzling Division, 4-5
Math + Science, A Solution, 5-9
Mostly Magnets, 2-8
Movie Math Mania, 6-9
Multiplication the Algebra Way, 6-8
Off the Wall Science, 3-9
Out of This World, 4-8
Paper Square Geometry:
    The Mathematics of Origami, 5-12
Puzzle Play, 4-8
Pieces and Patterns, 5-9
Popping With Power, 3-5
Positive vs. Negative, 6-9
Primarily Bears, K-6
Primarily Earth, K-3
Primarily Physics, K-3
Primarily Plants, K-3

Problem Solving: Just for the Fun of It! 4-9
Problem Solving: Just for the Fun of It! Book Two, 4-9
Proportional Reasoning, 6-9
Ray's Reflections, 4-8
Sensational Springtime, K-2
Sense-Able Science, K-1
Soap Films and Bubbles, 4-9
Solve It! K-1: Problem-Solving Strategies, K-1
Solve It! 2nd: Problem-Solving Strategies, 2
Solve It! 3rd: Problem-Solving Strategies, 3
Solve It! 4th: Problem-Solving Strategies, 4
Solve It! 5th: Problem-Solving Strategies, 5
Spatial Visualization, 4-9
Spills and Ripples, 5-12
Spring into Math and Science, K-1
The Amazing Circle, 4-9
The Budding Botanist, 3-6
The Sky's the Limit, 5-9
Through the Eyes of the Explorers, 5-9
Under Construction, K-2
Water Precious Water, 2-6
Weather Sense: Temperature, Air Pressure, and Wind, 4-5
Weather Sense: Moisture, 4-5
Winter Wonders, K-2

Spanish Supplements*
Fall Into Math and Science, K-1
Glide Into Winter with Math and Science, K-1
Mostly Magnets, 2-8
Pieces and Patterns, 5-9
Primarily Bears, K-6
Primarily Physics, K-3
Sense-Able Science, K-1
Spring Into Math and Science, K-1

* Spanish supplements are only available as downloads from the AIMS website. The supplements contain only the student pages in Spanish; you will need the English version of the book for the teacher's text.

Spanish Edition
Constructores II: Ingeniería Creativa Con Construcciones LEGO® 4-9
    The entire book is written in Spanish. English pages not included.

Other Publications
Historical Connections in Mathematics, Vol. I, 5-9
Historical Connections in Mathematics, Vol. II, 5-9
Historical Connections in Mathematics, Vol. III, 5-9
Mathematicians are People, Too
Mathematicians are People, Too, Vol. II
What's Next, Volume 1, 4-12
What's Next, Volume 2, 4-12
What's Next, Volume 3, 4-12

For further information write to:
AIMS Education Foundation • P.O. Box 8120 • Fresno, California 93747-8120
www.aimsedu.org • 559.255.6396 (fax) • 888.733.2467 (toll free)